# Simple Transistor Projects for Hobbyists & Students

No. 542
$7.95

# Simple Transistor Projects for Hobbyists & Students

**By Larry Steckler**

**TAB BOOKS**
Blue Ridge Summit, Pa. 17214

FIRST EDITION

FIRST PRINTING—DECEMBER 1970
SECOND PRINTING—SEPTEMBER 1972
THIRD PRINTING—NOVEMBER 1973
FOURTH PRINTING—FEBRUARY 1975

Copyright © by TAB BOOKS

Printed in the United States
of America

Reproduction or publication of the content in any manner, without express permission of the publisher, is prohibited. No liability is assumed with respect to the use of the information herein.

Hardbound Edition: International Standard Book No. 0-8306-0542-8

Paperbound Edition: International Standard Book No. 0-8306-9542-7

Library of Congress Card Number: 77-133800

# Preface

In this collection of circuits and projects there is literally something for everybody—student, experimenter, pure hobbyist, or anyone, really, with an interest in electronics. But you will find notable differences if you compare this book with most "project" books. Rather than follow the typical format of schematic, parts list, and a brief circuit description, we've placed more emphasis on how and why it works, particularly in the last two sections, so that those who are interested can better understand what they're doing. Of course, we have included schematics, and parts lists, too, when the component values and specifications do not appear on the schematic.

The first section or introduction is intended to acquaint you with some of the basic semiconductor construction techniques, if you've had little experience in that department. Then the next three sections offer a wide variety of projects, some quite simple and some more complex, to suit a variety of needs or desires.

The last two sections, which comprise a good portion of the book, are devoted to those fascinating new devices called Trigacs (Section 5) and a number of innovative integrated circuits (Section 6). If you want a better grasp on the operation of such semiconductors the information is there. If you simply want to build, you can skip most of the text material and go to work. At any rate we think you'll find the projects in this book quite intriguing, and at the same time they should help you keep pace with technology.

Larry Steckler

# Contents

## 1 INTRODUCTION 7

Modern Breadboards—Handling Semiconductors

## 2 PROJECTS FOR THE HOME & CAR 15

Sequential Turn-Signal System—Automatic Liquid-Level Control—High-Intensity Lamp Dimmer or Motor-Speed Controller—Tachometer for Optical or Electrical Sensing—Solid-State Control for Electric Blankets—Electronic Automobile Ignition System—Audio Controlled Lamp (Electronic Organ)—Full-Wave Heater Control—Universal Motor Speed Control with Built-in Timer

## 3 MORE PROJECTS FOR THE HOME & CAR 31

Emergency Lighting System—DC Flashers with SCRs—Proximity Switch or Touch Control—Slave Electronic Flash—Precision Two-step Thermostat—AC Power Flasher—DC Flashers—Automobile Burglar Alarm—Car Lights-on Alert—Four Motor-Speed Control Circuits

## 4 GADGETS FOR THE HOME 48

Electronic Bachelor Light—High-gain Limited-range Light Control—Isolated Low-voltage Remote Control—Frequency-selective AC Amplitude Control Circuit—15-ampere Battery Charging Regulator—3-position Power Control—High-Power Battery-operated Flasher with Photoelectric Control—Sequence Programmer and Power Driver—35-watt Audio Amplifier uses Plastic Transistors—30-watt 175-MHz Power Amplifier

## 5  TRIGAC — 62

Transfer Characteristics—The uA742 Circuit Outline—Trigac Operation from an AC Supply—Operation from a DC Supply—Hysteresis and Time Proportioning Operation—Applications of the Trigac—One Phase AC Circuits

## 6  PROJECTS USING ICs — 112

IC Regulated Power Supplies—1-watt Monolithic IC Amplifier—Using an IC Diode Array—Digital Organ Tone Generator—An IC Stereo Preamp

# SECTION I
# Introduction

Building and using electronic circuits is a practical and pleasurable hobby which is attracting more and more devotees every day. And we believe the circuits in this book will add to your enjoyment as well as provide many useful gadgets and devices. One of the best sources of circuits is the literature from semiconductor manufacturers. In fact most of those in this book were taken from material prepared by GE, RCA, Motorola, Texas Instruments and Fairchild. And to the above mentioned manufacturers we extend our appreciation for their splendid cooperation.

Of course, it is up to you to translate these circuits and descriptions into working assemblies. And to do so, I am sure you will use many different approaches. Some will use a standard chassis and wiring. Others will use a perforated board, flea clips, and point-to-point wiring as shown in Fig. 1-1. Still others will go to printed-circuit boards.

MODERN BREADBOARDS

If you're a breadboard enthusiast, there is now a new approach using .0035-inch fiberglass substrate and etched component pads. This completely modern technique eliminates the need to draw complicated artwork (a fact the non-artist should appreciate) and the intricate etching process required with printed boards. In Fig. 1-2 the hobbyist is placing component pads, called "Zaps," on a substrate board as he prepares to "grow" a circuit. The component pads adhere readily to almost any circuit, and once they're in place, you are ready to drill holes for the component leads (Fig. 1-3).

Fig. 1-1. Circuits can be breadboarded on a vectorboard. Here is a typical vectorboard arrangement on the bench.

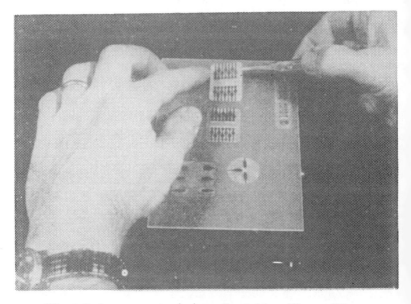

Fig. 1-2. An easy way to breadboard is with precision etched component pads on a 0.0035 fiberglass substrate. This technique eliminates the usual artwork and etch processes in making printed-circuit boards.

After drilling, the component connections are made by circuit "Zap" paths or "Zap" cords as you can see in Fig. 1-4. The completed circuit in Fig. 1-5 offers all the advantages of a printed board, plus the convenience and flexibility of a conventional breadboard. The experimenter's kit, developed by Bishop Graphics and available from GC Electronics (Fig. 1-6) contains enough material to build a number of circuit boards.

There is an even simpler breadboarding kit, called the "Mu-Dec," that requires no drilling and, of course, no soldering. The Type A general purpose kit shown in Fig. 1-7 is designed for integrated circuit use. The kit contains 45 component pads that simply plug into appropriate holes on the boards, with four attachments on each pad. The plug-ins are designed to accommodate flat-pack ICs or the standard TO-5 type components.

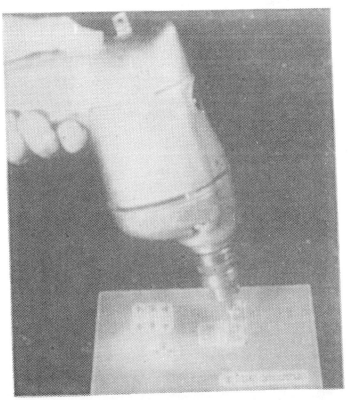

Fig. 1-3. Circuit "Zaps" are self-adhering to almost any surface and need only be drilled to "Zap-in" circuit connecting terminals.

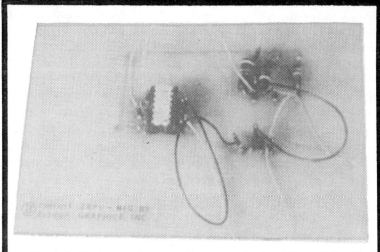

Fig. 1-4. Component connections may be by "Circuit Zap" paths or the very convenient "Zap" cords.

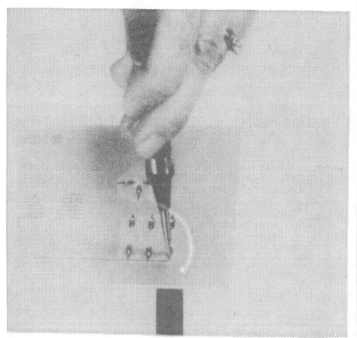

Fig. 1-5. Completed test circuit has the convenience of a printed-circuit board, yet retains the flexibility of change so common with breadboard circuits.

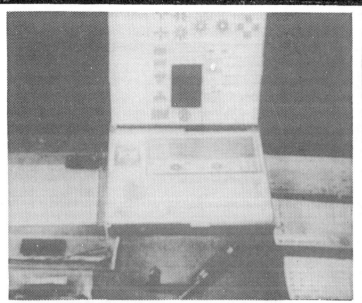

Fig. 1-6. Bishop Graphic's Circuit Zap kit provides instant printed-circuits for everyone.

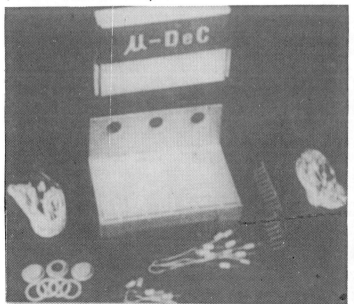

Fig. 1-7. The British contribution to rapid circuit breadboarding is the solderless modular breadboard. No drilling, no soldering, no etching. Just push in component leads and interconnect with teflon coated jumper leads.

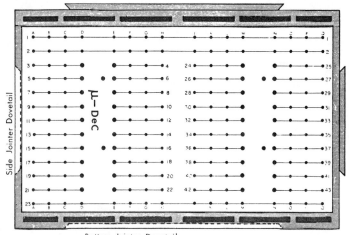

Fig. 1-8. The Mu-Dec Type "A" general purpose breadboard for integrated circuits features modular attachment by dovetail interlocks as well as 45 component pads that provide four attachments on each pad. Plug-in sockets for flat-pack integrate circuits or the standard TO-5 sockets simplify solid-state breadboarding.

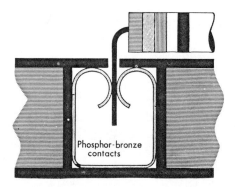

Fig. 1-9. Hard tempered phosphor-bronze contacts provide reliable and long-life connection with components or leads.

The drawing in Fig. 1-8 shows the connections available on the board. The larger holes (dots on the drawing) accept the plug-in pads and the numerous smaller holes are used for components and interconnecting patch leads. The contacts on the board are made of phosphor-bronze and are designed for repeated use (Fig. 1-9). A panel to mount potentiometers and other controls is included in the kit.

When using a printed board, don't expect to mount large, heavy, or hot components on the board itself. This is highly impractical and not what the board is intended for. Large transformers, bulky controls, high-wattage resistors, and power transistors should be mounted on a sturdy metal chassis for proper support and improved heat dissipation—not on a circuit board.

HANDLING SEMICONDUCTORS

In spite of the rugged durability of solid-state equipment, semiconductor devices, individually and out of the circuit, are sensitive to extreme heat, mechanical vibration and electric shocks that exceed voltage ratings. Once semiconductors are assembled in a circuit, provided they are installed carefully and correctly, they are protected from heat and shock (unless something goes wrong in the circuit).

Therefore, it is important that you handle semiconductors correctly during assembly. When forming the leads to mount a transistor, IC, diode, etc., don't bend them too close to the case. Also, when bending a lead it's a good idea to grip it with long-nose pliers close to the case to protect the semiconductor against strain and vibration. The same applies when you trim leads; use the long-nose pliers to hold the lead at a point between the cut and the case. Deliberate caution when working with semiconductors will help prevent disappointing failures.

PC DO'S & DON'TS

Be careful not to damage the copper foil by overheating; it's best to use a pencil-type soldering iron and apply just enough heat to make a good connection. Overheating can inflict irreparable damage on a PC board!

Do be careful when removing components from a PC board. Make sure the solder has melted before trying to pull leads from the mounting holes. Unnecessary force can damage the etching.

Do not apply excessive pressure when removing any component. Never use a screwdriver to pry around on a PC board.

Do not use an excessive amount of solder. It can very easily run onto adjacent areas of the etching and cause short circuits.

Do not use soldering paste on PC wiring. Also, be sure to clean the board thoroughly, removing excess solder or little slivers of solder.

Do not forget to tie down damaged circuit etching or remove the bad section and replace with wire.

# SECTION II

# Projects for the Home & Car

SEQUENTIAL TURN-SIGNAL SYSTEM

Here is a device suitable for use with any automobile with three rear lights on each side (such cars include many Cheverolet models since 1958, late model Buicks, 1966 Continentals, some Mercurys, etc.). What it amounts to is a solid-state sequential turn signal simulating a flashing arrow at the rear of the car by turning on the rear lights in sequence. Since this is an all-electronic system it requires no relays. Fig. 2-1 shows the lamp pattern. In addition, the same circuit can be used for barricade lights, beacons, portable advertising signs, etc. Additional stages can be added as desired. However, if you do add stages you will have to "size" the thermal flasher module for the total number of lamps that are to be driven.

When the turn signal switch is operated, voltage is applied to the appropriate flasher circuit (one circuit is used for the left-hand bank of lights, and another, duplicate circuit for the

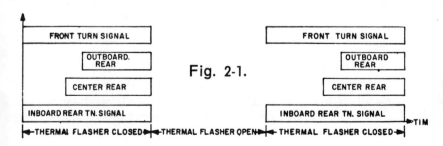

Fig. 2-1.

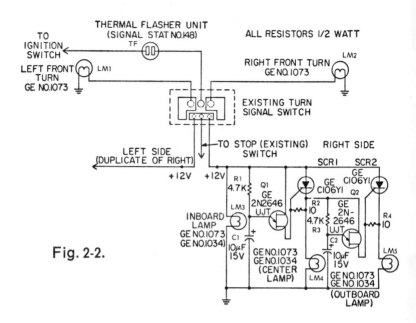

Fig. 2-2.

right-hand bank) through a standard thermal flasher module (get the flasher modules from an automotive supply house). Since the inboard lamp and the front lamp are connected directly to the output of the thermal flasher, both lamps light at the same time. Simultaneously, voltage is applied to the unijunction transistor time-delay circuit connected across the inboard lamp. After about 0.25 second, an output pulse from the unijunction timer triggers SCR1 (see Fig. 2-2) into conduction and SCR1 energizes the center lamp. A second unijunction timer connected across the center lamp is then energized, and after another 0.25 second or so, its output triggers SCR2 into conduction. SCR2 then energizes the outboard lamp. After a period with all three lamps on, the thermal flasher opens, power is removed from the circuit and all the lamps go out. The sequence repeats each time the thermal flasher recycles as long as the turn indicator is on.

Installing the device in the car is easy and straightforward. The two leads (in the car trunk) going to the existing left and right flasher lamps are cut at some convenient point and the "hot" end of each lead is connected to the positive 12-volt terminal of an electronic flasher unit. The three lamp outputs from each flasher unit are then connected to the appropriate rear lamps. In those cars where one of the rear lamps was originally used as a back-up light, disconnect its lead wire, tape it up, or connect it to a separate backup light. Fit a red lens to the old backup light. To complete the installation, change the existing thermal flasher unit to the one shown in the schematic (Fig. 2-2).

**Parts List (Fig. 2-2)**

R1, R3: 4,700 ohms, 1/2 watt
R2, R4: 10 ohms, 1/2 watt
C1, C2: 10 mfd, 15v electrolytic
LM1, LM2: GE 1073 lamp
LM3, LM4, LM5: GE 1073 or 1034 lamp
Q1, Q2: GE 2N2646 unijunction
SCR1, SCR2: GE C106Y1
TF: Thermal flasher, Signal Stat 148

AUTOMATIC LIQUID-LEVEL CONTROL

Simple and highly effective, this low-cost control is designed to maintain the liquid content of any reasonably well-grounded

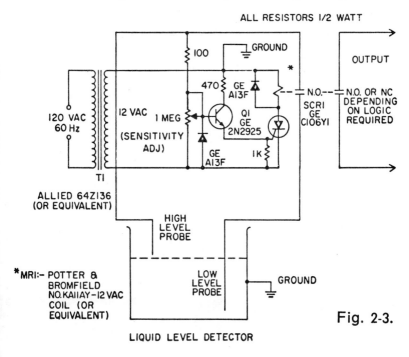

Fig. 2-3.

LIQUID LEVEL DETECTOR

container between two previously established limits. Liquid level is detected by two metal probes; one measuring high level, the other low level. Both probes are of single-wire construction to optimize limiting effects and both are energized with alternating current to avoid electrolytic corrosion. A sensitivity adjustment is provided so the circuit can accommodate liquids of different conductivities.

To see how the circuit operates, look at Fig. 2-3. With the liquid level just below the low-level probe, no base current flows to transistor Q1, the SCR is off, and output relay RY1 is de-energized. Should the liquid level rise above the low-level probe, the circuit state remains unchanged, since there is a normally-open relay contact in series with the probe. When the liquid level reaches the high-level probe, however, base current flows to Q1 (via ground, the liquid, and the probe), Q1 triggers SCR1 and SCR1 energizes relay RY1. Once the relay is energized, the normally-open contact in series with the low-level probe closes and latches the circuit on until the liquid level falls below the low probe. Diode D1 removes reverse voltage from Q1's base-emitter junction during negative half cycles of the AC supply and also insures that AC current flows through the probes to prevent electrolytic corrosion.

The controlled device (pump, solenoid, etc.) can be taken directly from the second set of relay contacts, which can be either normally-open or normally-closed, depending upon the logic required. Alternately, the relay contacts can be used to drive a Triac AC power switch. In the latter case, the Triac handles the high-power load, while the relay contacts supply gate current to the Triac. Since the Triac gate pre-

**Parts Lists (Fig. 2-3)**

R1: 100 ohms, 1/2 watt
R2: 1 megohm potentiometer
R3: 470 ohms, 1/2 watt
R4: 1000 ohms, 1/2 watt
D1, D2: GE A13F
RY1: See note in diagram
SCR1: GE C106Y1
Q1: GE 2N2925
T1: Filament transformer: Primary 120v; secondary 12v (Allied 64Z136)

sents a low resistive load to the contacts, arcing, burning and wear problems are reduced.

## HIGH-INTENSITY LAMP DIMMER OR MOTOR-SPEED CONTROLLER

Operating on the phase-control principle, this low-cost variable voltage control was originally designed to adjust the speed of small 120-volt AC shaded-pole fan motors with current ratings up to 1.5 amperes. The speed range provided by the control is approximately 3:1. With slight modifications, this circuit can also be used to dim high-intensity, low-voltage table-top lamps: (NOTE: THIS CIRCUIT MUST NOT BE USED FOR DIMMING 120-VOLT INCANDESCENT LAMPS.)

So that the single GE C106B SCR can exert control over both half-cycles of the AC supply, a diode bridge is used to convert AC into full-wave pulsating DC which is applied directly to the SCR. The AC load (the motor or the primary winding of the lamp step-down transformer) is connected in series with one leg of the AC supply. Because the GE C106 SCR can be triggered with less than 0.5 ma gate current, a type 5AB neon

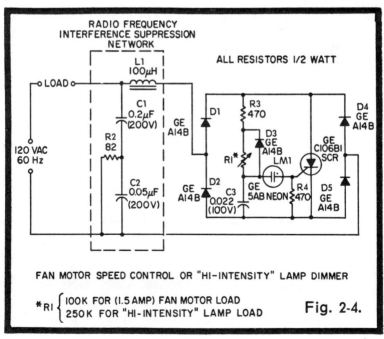

FAN MOTOR SPEED CONTROL OR "HI-INTENSITY" LAMP DIMMER

*R1 { 100K FOR (1.5 AMP) FAN MOTOR LOAD
      250K FOR "HI-INTENSITY" LAMP LOAD

Fig. 2-4.

glow lamp makes an ideal trigger device. The voltage-adjustment potentiometer R1 and the 0.022-mfd timing capacitor in conjunction with the neon glow lamp form a relaxation oscillator whose frequency determines the SCR trigger point in each half cycle.

The GE A14B rectifier connected across the potentiometer resets the timing capacitor to zero volts each time the SCR triggers and insures that the output waveform is symmetrical from one half cycle to the next (that is, no DC component in the load voltage). For use as a lamp dimmer, change the value of the 100K potentiometer to 250K. The RF interference filter must not be omitted, because in addition to acting as a noise filter it prevents fast-rising transients from inadvertently triggering the SCR.

**Parts List (Fig. 2-4)**

R1: potentiometer 100K for the fan-motor control
250K for the high-intensity lamp control
R2: 82 ohms, 1/2 watt
R3, R4: 470 ohms, 1/2 watt
C1: 0.2 mfd, 200 volts
C2: 0.05 mfd, 200 volts
C3: 0.022 mfd, 100 volts
L1: 100 microhenry choke
LM1: GE 5AB neon lamp
D1, D2, D3, D4, D5: GE A14B
SCR: GE C106B1


TACHOMETER FOR OPTICAL OR ELECTRICAL SENSING

This circuit uses a light-activated SCR (LASCR) as a low-current gate turn-off switch. After being triggered by light or by an electrical pulse, the LASCR will continue to conduct until the current in the gate inductance builds up to a value large enough to turn it off. Since the current through the meter continues for a fixed length of time after each triggering pulse, the average current (and, therefore, the meter reading) will depend on the repetition rate of the triggering pulses.

If the interval between trigger pulses is say 10 percent longer than the on time of the SCR, the average current through the SCR will be 90 percent of the peak current. Set the voltmeter so this condition represents full-scale meter deflection. A further decrease in time between trigger pulses now results in a pulse-skipping or frequency-dividing action in which the SCR conducts only on alternate pulses. This causes the meter indication to drop to half scale, automatically providing a range change of 2:1.

The optical tachometer shown in Fig. 2-5 is an example of the use of a constant-pulse-width circuit. The average value of anode current in the LASCR is directly related to the repetition rate of the trigger pulses, either light or electrical. The voltmeter is calibrated in RPM and essentially measures the average current in the LASCR. Since the pulse width is determined by both anode current, peak amplitude, and by the gate inductance, the simplest method of calibrating this tacho-

Parts List (Fig. 2-5)

R1: 470 ohms, 1/2 watt
R2: 5K pot
R3: 10K potentiometer
D1: Zener diode, GE Z4XL9.1
D2: GE 1N1692
L: 1-henry choke
LASCR: GE L9U
M: 0-6v DC, 200 ohms per volt meter

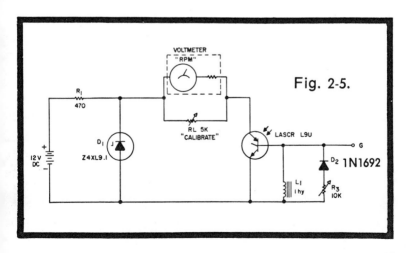

Fig. 2-5.

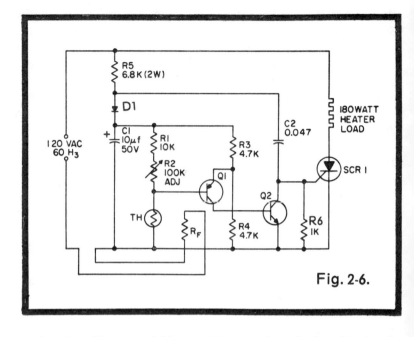

Fig. 2-6.

meter is with a variable resistor, R2, which adjusts the pulse width by the anode current peak. The damping resistance R3, should be set for the maximum value which will permit one-shot operation of the circuit. If R3 is too large, the circuit will oscillate by itself. If R3 is too small, current through L1 will not decay rapidly and there will be a significant period of time immediately after turn-off when the LASCR cannot be retriggered.

Electrical signals can also be used to drive this tachometer by coupling pulses into the gate terminal of the LASCR through a high enough impedance to prevent interference with the conduction pulse width.

SOLID-STATE CONTROL FOR ELECTRIC BLANKETS

This low-cost solid-state temperature controller duplicates the function of the bimetal switch normally supplied with electric blankets. It is equally effective for use with other low-power heating loads as long as they are 180 watts or less. Because the circuit works on the "integral cycle" principle (only complete half cycles of power are applied to the load), radio-frequency interference is minimized without the need for expensive and bulky filter circuits.

The circuit shown in Fig. 2-6 works as follows. Rectifier D1 and capacitor C1 form a simple DC power supply for the 4-arm resistance bridge consisting of R1 and R2, the thermistor, R3 and R4. The base-emitter input terminals of transistor Q1 are connected across the output terminals of this bridge, and Q1 amplifies bridge unbalance, if any. When the bridge is balanced (adjustable with potentiometer R2) there is no output and Q1 is cut off. Therefore, transistor Q2 receives no base drive. During the initial portion of each positive half cycle of the AC supply, before diode D1 starts to conduct, capacitor C2 couples current into the gate of SCR1. Then SCR1 triggers and energizes the load. Notice that once D1 starts to conduct, capacitor C2 is clamped to a DC voltage established by capacitor C1 and gate current stops flowing. Thus, SCR1 can trigger only at the start of each positive half cycle. In this way radio-frequency interference is minimized.

With current flowing through the heater load, feedback resistor Rf heats and raises the temperature of the monitoring thermistor. As the thermistor's resistance drops, the bridge unbalances and transistor Q1 starts to conduct. Transistor Q2 receives base drive from Q1; it is driven into saturation and shunts gate current away from SCR1. Then SCR1 turns off and removes power from the load. SCR1 comes back on again only when the monitoring thermistor cools to room ambient.

### Parts List (Fig. 2-6)

R1: 10K
R2: 100K potentiometer
R3, R4: 4.7K
R5: 6.8K, 2 watts
R6: 1K
Rf: see text
C1: 10 mfd, 50 volts, electrolytic
C2: 0.047 mfd
TH1: Thermistor, 50K NTC
D1: GE A13C
Q1: 2N3638
Q2: GE 2N3415
SCR1: GE C106B1

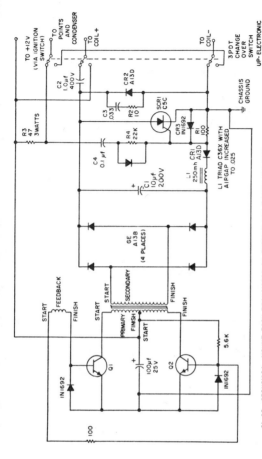

Fig. 2-7.

## Parts List (Fig. 2-7)

R1, R2: 33 ohms, 1/2 watt
R3: 10 ohms, 1 watt
R4: 47 ohms, 2 watts
R5: 33K, 1/2 watt
R6: 330 ohms, 1/2 watt
R7: 100 ohms, 1/2 watt
R8: 10 ohms, 1/2 watt
R9: 56K, 1/2 watt
C1: 10 mfd, 200 volts, electrolytic
C2: 0.033 mfd, disc ceramic
C3: 1 mfd, 400 volts, molded
C4: 0.1 mfd, molded
D1, D7, D9: GE A13F
D2, D3, D4, D5: GE A13B
D6, D8: GE A13C
Q1, Q2: GE 11C1536

SCR: GE C6C
L1: 300 mh, 100 ma.
T1: see the note in Fig. 2-7.

The on-off cycle repeats itself at a rate determined by room temperature, with the on time increasing if room temperature falls and decreasing if the room heats up.

## ELECTRONIC AUTOMOBILE IGNITION SYSTEM

The system described here is an SCR capacitor-discharge unit driven by a silicon-transistor inverter. By using all-silicon semiconductors, the unit can provide reliable operation even at high ambient temperatures. Germanium semiconductors cannot operate safely at high temperatures.

Battery drain is low, less than two amperes, over the entire speed range. The circuit will provide full spark voltage up to 6000 RPM on an 8-cylinder engine. Load on the distributor points is very light; so there is no arcing and very little wear. In most instances the rubbing block or spring wears before the point surfaces.

To simplify its use, the existing points and capacitor, as well as the existing ballast resistor and coil, are used. A lock-out circuit in the gate of the SCR prevents false triggering caused by point bounce, thus eliminating misfiring. Where desired the contact points can be eliminated entirely and the SCR triggered by the pulse output from a magnetic pickup attached to the distributor.

Let's look at the circuit action (Fig. 2-7). Transistors Q1 and Q2 in combination with transformer T1 form a very efficient saturating core square-wave inverter that delivers over 160 volts DC into filter capacitor C1. The discharge capacitor (C2) charges to double this voltage through the resonant charging action of C2, choke L1 and diode D1. When the car's distributor points open, SCR1 is triggered by current flowing from the battery to charge capacitor C4. At this point capacitor C2 is connected across the coil's primary winding and a high voltage is induced in the coil secondary by transformer action. This high-voltage pulse is fed to the appropriate spark plug by the regular car distributor.

Because capacitor C2 and the primary inductance of the ignition coil form a second oscillatory circuit, capacitor C2 overswings in voltage and this reverse voltage turns off the SCR. Any excess energy remaining in C2 is then fed back to charge C2 in the forward direction via bypass diode D2. Re-

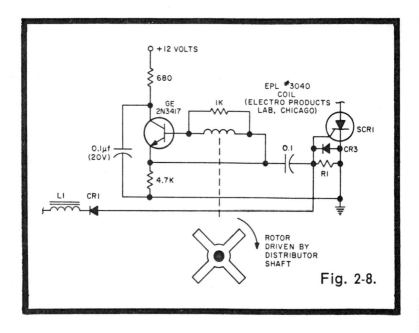

Fig. 2-8.

sistor R2 and capacitor C3 limit the rate of forward voltage across the SCR within its DV/DT ratings. When the breaker points close once more, capacitor C4 is discharged in readiness for the next cycle through R4. A relatively long time-constant is provided for C4 to lessen the false triggering by providing a negative gate bias to SCR1 whenever charging current is flowing from the DC supply to charge capacitor C2. This circuit will deliver approximately 23,000 volts output from a 12.6-volt car battery input and will continue to operate at battery voltages as low as 7.0 volts. A contactless triggering circuit that can be used with this system is shown in Fig. 2-8.

## AUDIO CONTROLLED LAMP (ELECTRONIC ORGAN)

If you want light that dances in step with your favorite music, here is the circuit that will turn the trick. It is an on-off control with an isolated, low-voltage input. Since the switching action is very rapid, compared to the response time of the lamp and the response of the human eye, the effect produced by this circuit is dramatic and interesting. As a heavy-duty Triac is used in the control circuit, incandescent lamps with ratings up to 500 watts can be controlled. Use a number of these circuits in conjunction with narrow-range audio filters and you can have different color lamps reacting to the various frequencies in the music controlling the lights. The circuit is in Fig. 2-9.

**Part List (Fig. 2-9)**

R1: 1K, 1/2 watt
R2: 100-ohm, 2-watt potentiometer
SCR: GE C6U
T1: Filament transformer: Primary 120 volts; Secondary 6.3 volts
Triac 1: GE SC41B

FULL-WAVE HEATER CONTROL

Triac circuits are the ultimate in simplicity for manual control of load power. This circuit (Fig. 2-10) is ideal for controlling heaters, fans, and over a limited range, even lamps (the SCR version of this circuit is shown in Fig. 2-11). To turn the circuit in Fig. 2-10 into a photo-electric control circuit, replace resistor R1 with a cadmium sulphide photocell

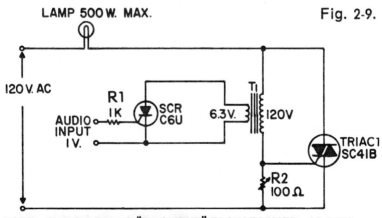

Fig. 2-9.

NOTE: $T_1$ IS A 6.3V, 1A. "FILAMENT" TRANSFORMER. ADJUST $R_1$ FOR MAXIMUM RESISTANCE THAT WILL NOT TURN ON LAMP WITH ZERO INPUT.

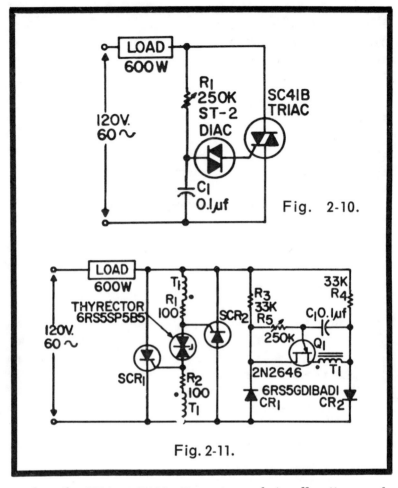

Fig. 2-10.

Fig. 2-11.

such as the GE type B425. To reverse photocell action, make resistor R1 10K and place the photocell in parallel with capacitor C1.

UNIVERSAL MOTOR SPEED CONTROL WITH
BUILT-IN TIMER

Particularly suitable for blenders and larger pedestal-type food blenders, this universal motor-speed control has a built-in electronic timer that shuts the motor off at the end of a predetermined time cycle. Both motor-speed and time-delay can be adjusted separately.

The motor control circuit has speed-dependent feedback that

gives excellent torque characteristics to the motor, especially at low rotational speeds where conventional rheostat or variable transformer controls are ineffective. Speed is adjusted with potentiometer R2. With the components shown, the control can handle motors rated up to 2 amperes. If a higher current rating is required, use a GE C23B SCR instead of the C22B unit.

The automatic shutoff feature is derived from circuitry associated with the C106B2 timer SCR. The timer SCR diverts gate current away from the main speed control SCR upon completion of a time cycle. Operation is as follows:

With the main power switch off, the motor is disconnected from the AC line, but capacitor C1 charges to the peak of the line through rectifier diode D1 and resistor R1. When the motor is turned on with the power switch, the pilot lamp lights and capacitor C1 starts to discharge through resistors R3 and rectifier D2. Since the time-constant associated with this RC network is numerically long and discharge current can flow only for a short period each cycle, capacitor C1 takes many complete cycles of AC to discharge. When C1 is finally discharged and starts to charge in the opposite direction, the timer SCR triggers and diverts gate current away from the main SCR. At this point the motor stops. For safety

**Parts List (Fig. 2-10)**

R1: 250K potentiometer,
C1: 0.1 mfd, paper
Triac: GE SC41B
Diac: GE ST-2

**Parts List (Fig. 2-11)**

R1, R2: 100 ohms. 1/2 watt
R3, R4: 33K, 1/2 watt
R5: 250K
C1: 0.1 mfd, paper
D1, D2: GE 6RS5GD1BAD1
T1: Pulse transformer, Sprague 35ZM923
Q1: GE 2N2646
Thyrector: GE 6RS5SP5B5

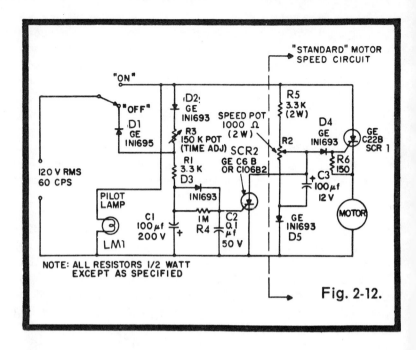

Fig. 2-12.

**Parts List (Fig. 2-12)**

R1: 3.3K, 1/2 watt
R2: 1K, 2-watt potentiometer
R3: 150K potentiometer
R4: 1 megohm, 1/2 watt
R5: 3.3K, 2 watts
R6: 150 ohms, 1/2 watt
C1: 100 mfd, 200 volts, electrolytic
C2: 0.1 mfd, 50 volts
C3: 100 mfd, 12 volts, electrolytic
LM1: 120-volt pilot lamp
D1: GE 1N1695
D2, D3, D4, D5: GE 1N1693
SCR1: GE C22B
SCR2: GE C6B or C106B2

the pilot lamp remains on until the main power switch is turned off. Maximum time delay is somewhat more than 30 seconds.

# SECTION III
# More Projects for the Home & Car

EMERGENCY LIGHTING SYSTEM

The circuit in Fig. 3-1 provides battery-operated emergency lighting instantaneously and automatically upon failure of the regular 120-volt line power. When normal power is restored, the emergency light goes out and the battery is recharged. The circuit is suitable for use in any location where the loss of normal lighting is undesirable, even for short periods of time.

Then the AC power is on, capacitor C1 charges with the polarity shown through diode D2 and resistor R2 and discharges through resistor R1 and the battery. Because the discharge time constant is longer than the charging time constant, capacitor C1 retains at all times a net positive charge as shown in Fig. 3-1. Under these conditions the SCR's gate is back biased and it cannot trigger. The battery is charged and kept charged through diode D1.

Should the AC power fail, capacitor C1 discharges completely and then starts to charge in the opposite direction, with power furnished by the battery. When voltage on capacitor C1 is large enough to trigger SCR1, it turns on and energizes the emergency light. Reset is automatic when AC power is restored, because SCR1 is reverse biased when D1 conducts to charge the battery.

If you replace the emergency lamp with a battery-operated bell or siren, you will have a simple power failure alarm. Use it to protect your freezer or heating plant against unexpected interruption in the AC power supply.

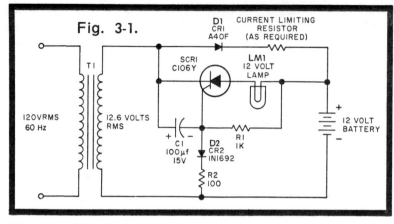

Fig. 3-1.

**Parts List (Fig. 3-1)**

R1: 1K  
R2: 100 ohms  
C1: 100 mfd, 15 volts, electrolytic  
D1: GE A40F  
D2: GE 1N1692  
LM1: 12-volt lamp  
SCR1: GE C106Y  
T1: Filament transformer: Primary 117v AC;   Secondary 12.6 volts

## DC FLASHERS WITH SCRs

A DC flasher is nothing more than an SCR flip-flop. The circuit shown in Fig. 3-2 has a variable on-off adjustment as an added convenience. The arrangement of diodes D1, D2, D3, and D4 makes it possible to independently adjust both on and off times of the load. The circuit itself is a capacitor commutated SCR flip-flop.

The SCRs conduct alternately and are triggered by the pulses out of base 1 of unijunction Q1. In this type of circuit it is important that at the start, when power is first applied to the circuit, that some method is provided to insure the triggering of only one SCR. The network consisting of R9, C3, and D5 takes care of this situation. When power is first applied, both SCRs are off. Because of the positive potential on the anode of SCR1, resistor R9 will apply the same potential to the cathode of D5, thus reverse biasing it. When a pulse appears at base 1 of Q1, only the gate of SCR1 will receive this pulse and only SCR1 will be turned on.

Capacitor C4 will now charge through Load 1 with a positive charge on the side connected to the anode of SCR2 and nearly

**Parts List (Fig. 3-2)**

R1, R2: 500K linear potentiometers
R3, R4: 750K, 1/2 watt
R5: 100 ohms, 1/2 watt
R6, R7: 1K, 1/2 watt
R8: 270 ohms, 1/2 watt
R9: 4.7K, 1/2 watt
R10: 250 ohms, 5 watts
C1: 0.47 mfd, 50 volts
C2, C3: 0.22 mfd, 50 volts
C4: 4 mfd, 50 volts, non-polarized electrolytic
Q1: GE 2N2646 unijunction
SCR1, SCR2: GE C106F
D1, D2, D3, D4, D5: GE A13F
Load: GE 50C, 1.4 ampere lamp

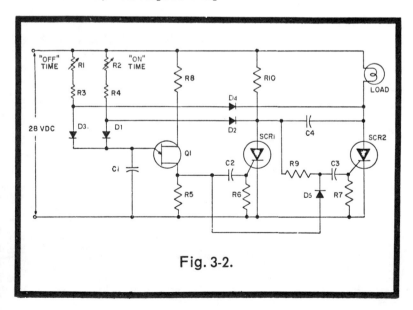

Fig. 3-2.

a ground potential at the anode of SCR1. With SCR1 on, the bias on D5 is reversed and the junction of D1 and D2 is clamped to nearly 1 volt because D2 is now forward biased. Capacitor C1 now starts charging through R1, R3 and D3. At the end of the time delay (may be adjusted by the setting of R1), the unijunction will produce another pulse, turning SCR2 on. This corresponds to connecting C4 across SCR1 so that it is momentarily reverse biased. The momentary reversal of anode potential turns SCR1 off. With SCR2 on, diode D4 is forward biased and, therefore, capacitor C1 starts charging through R2, R4, and D1. With this arrangement, therefore, the off time (SCR2 off) is determined by the setting of R1 and the on time is determined with the setting of R2.

SCR2 should be selected so that the maximum load current is within its rating. Since SCR1 is used for commutating SCR2, it can have a smaller rating that SCR2. Notice that the greater the current through the load, the larger the value capacitor C4 must have. The minimum required value for C4 can be easily determined by this equation:

$$C4 \geq \frac{1.5 t_{off} I}{E}$$

Where C4 is in mfd; $t_{off}$ equals the turn-off time of the SCR in microseconds; I is the maximum load current (including possible overloads) in amperes at the time of commutation; E represents the minimum DC supply voltage.

If the anode of SCR1 has a lamp in its anode as a load, rather than resistor R10, the circuit as shown would not function properly because when SCR1 is on, the trigger pulse is coupled to both gates and SCR1 would not have enough time to turn off due to the short time constant involved. With the component values shown, however, triggering SCR1 and SCR2 at the same time is not objectionable because the time constant R10-C4 is much longer than the trigger-pulse width so that SCR1 remains reverse biased after the end of the trigger pulse long enough to assure commutation of SCR1. To be able to drive equal loads in the anodes of the SCRs, the SCR1 gate would need a bias similar to the SCR2 gate bias. If this is the case, some additional starting means would have to be incorporated into the circuit.

## PROXIMITY SWITCH OR TOUCH CONTROL

Here is a solid-state version of the well-known vacuum-tube operated proximity switch. It is useful for such applications as elevator touch buttons, door control, light switches, etc.

In Fig. 3-3 capacitor C1 and the sensor plate ("capacitor" C2) form a capacitive voltage divider connected directly across the AC supply. The AC voltage across C1 will depend upon

**Parts List (Fig. 3-3)**

R1, R2: 100K
R3: 39K
C1: 10 pf
C2: Touch plate capacitance
D1: GE 1N1692
LM1: NE-83 neon lamp
Relay: 12-volt unit
SCR: GE C7U or C106
T1: Autotransformer: Primary, 120 volts; Secondary, 12.6-volt tap

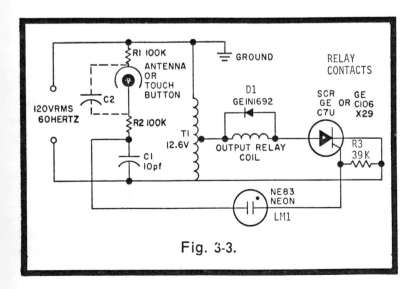

Fig. 3-3.

the ratio C1 divided by C2 and the line voltage. The capacitance of C2, in turn, depends on the proximity to the sensor plate of any reasonably conductive and grounded object (metals, human body, etc.). As soon as the voltage across C1 exceeds the breakover potential of the NE-83 neon lamp, capacitors C1 and C2 discharge through the SCR gate, causing the SCR to trigger and energize the load. Resistors R1 and R2 limit the current through the touch button to a safe level.

Latching action can be provided, if desired, by driving the SCR <u>anode</u> <u>circuit</u> <u>only</u> with DC. Since the sensitivity of the arrangement is a function of both sensing distance and sensor plate size, the plate can be made progressively smaller in area if the sensing distance is small. For a touch control, the sensing plate need be no larger than a penny.

SLAVE ELECTRONIC FLASH

In photography today, there is much demand for a flash photosensitive switch capable of triggering the "slave" flash units used extensively in multiple-light-source high-speed photography. The schematic in Fig. 3-4 shows how a standard electronic flashgun circuit can be modified with a light-activated SCR (LASCR) to serve as a fast-acting slave flash.

With switch S1 closed, capacitor C1 charges to 300 volts through resistor R1 and capacitor C2 charges to approximately 200 volts through resistors R2 and R3. When the master flashgun fires (triggered by the flash contacts on the camera), the light output triggers LASCR1 which then discharges capacitor C2 into the primary winding of transformer T1. Its secondary puts out a high-voltage pulse to trigger the flashtube. The flashtube discharges capacitor C1, while the resonant action between C2 and T1 produces a voltage that reverse-biases LASCR1 for positive turn off. With the intense instantaneous light energy available from present-day electronic flash units, the speed of response of the LASCR is easily in the low microsecond region, producing perfect synchronization between master and slave.

High levels of ambient light can also trigger the LASCR when a resistor is used between gate and cathode. Although this resistance could be made adjustable to compensate for ambient light, the best solution is to add an inductance (at least 1 henry) which will appear as a low impedance to ambient light and as a very high impedance to a flash.

## PRECISION TWO-STEP THERMOSTAT

The elementary bimetal thermostat is subject to problems of mechanical loading, slow make and break of contacts, and self heating with current flow through it. Two light-activated SCRs, a lamp, and a bimetal element can provide precise temperature regulation with a two-step power control. See Fig. 3-5.

As temperature increases, the bimetal element blocks light from LASCR1, reducing the heater to half power. Then a

**Parts List (Fig. 3-4)**

R1: 810 ohms, 2 watts
R2: 1 megohm
R3: 1.8 megohm
R4: 56K
C1: 1000 mfd, 400 volts, electrolytic
C2: 0.22 mfd
L1: 1 henry choke
LM1: GE FT 106 flashlamp
LASCR: GE L8B
S: SPST toggle switch
T1: UTC PF7 transformer

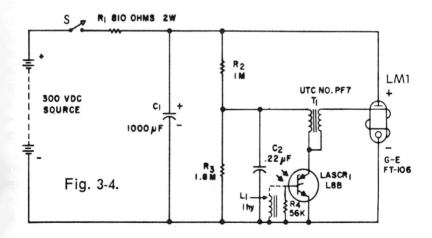

Fig. 3-4.

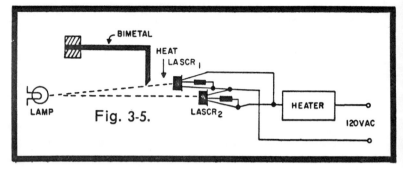

Fig. 3-5.

further increase in temperature causes the bimetal element to block light from LASCR2, turning the heater off completely. Since there is no mechanical loading, the differential of this thermostat is very small and is determined primarily by optics and the change in light sensitivity of the LASCR with temperature and voltage.

AC POWER FLASHER

Most flashers available today have a motor-driven cam that opens and closes heavy silver contacts. The arc generated the instant the contacts open and close, the high inrush current occurring when a tungsten lamp is switched on, and the mechanical wear of the contacts limit the operating life of this system.

The circuit in Fig. 3-6 illustrates a basic AC flasher that has no moving parts. It is basically a free-running unijunction

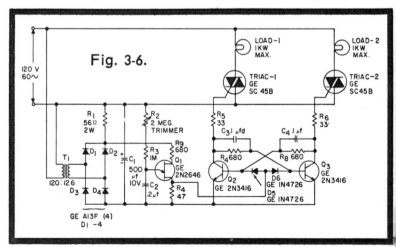

Fig. 3-6.

oscillator triggering a transistor flip-flop, which in turn alternately fires two Triacs capable of handling a 1-kw load each. If a single lamp output with only on-off performance rather than two alternately flashing lamps is desired, leave out Triac 2, but be sure to make the proper compensating connection (see Fig. 3-6).

Here's how the circuit works: Transformer T1, diodes D1, D2, D3, D4, resistor R1 and capacitor C1 provide the DC supply to the free-running unijunction oscillator Q1 and to the transistor flip-flop Q2-Q3. Because of the ripple on base 2 of unijunction Q1, capacitor C2 can reach the peak-point voltage of Q1 only at the beginning of the half cycles, thus firing Q1 early in the half cycle. The synchronization of unijunction Q1 minimizes the effect of radio-frequency interference. The oscillation frequency of the Q1 circuit is determined by the setting of R2.

Collector-gate resistors R5 and R6 form a divider network, with R1 supplying about 6 volts DC to the flip-flop. Suppose, initially, that Q2 is on and Q3 is off. In this case, the collector of Q2 is a negative potential with respect to the gate and Terminal 1 of Triac 1, while the collector of Q3 is at the same potential as the gate and Terminal 1 of Triac 2.

The negative potential seen at the gate causes current to flow out the gate from the positive side of the DC supply, through Terminal 1 and the gate of Triac 1, R5, collector Q2 and out of the emitter of Q1 to the negative side of the DC supply. Current flow out of the gate of Triac 1 causes it to conduct, energizing Load 1. Since the gate and Terminal 1 of Triac 2 do not see a different potential, there is no current flow to or from the gate and, therefore, Triac 2 remains off.

Timing capacitor C2 charges through R2 and R3, and when the voltage across it reaches the peak point voltage of unijunction transistor Q1, it discharges, producing a negative-going pulse across resistor R4. A negative-going pulse at the junction of C3 and C4 changes the state of the flip-flop, turning Q2 off and Q3 on, causing Triac 1 to stop conducting and Triac 2 to conduct. In this manner, the Triacs turn on and off alternately every time the unijunction fires. It should be noted that the on time is equal to the off time with the unijunction connected as shown in Fig. 3-6. This does not permit the variation of only one of the timings without changing the other one as well. To get independent timing for the on and

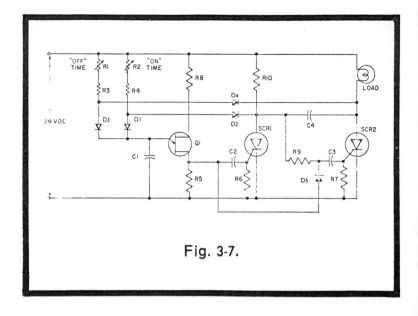

Fig. 3-7.

off functions, diode gating is required. (The component values are shown in Fig. 3-6.)

## DC FLASHERS

DC flashers are nothing more than an SCR flip-flop. The schematic in Fig. 3-7 shows such a circuit and includes variable on-off adjustments. The arrangement of diodes D1, D2, D3, and D4 makes it possible to independently adjust both on and off times of the load. The circuit is a capacitor-commutated SCR flip-flop.

The SCRs conduct alternately and are triggered by the pulses out of base 1 of unijunction Q1. In this type of circuit it is important, at the start, when power is first applied to the circuit, that some means is provided to insure the triggering of only one SCR. The network of R9, C3, and D5 takes care of this situation.

When power is applied, both SCRs are off. Because of the positive potential on the anode of SCR1, the same potential is applied (through resistor R9) to the cathode of diode D5, thus reverse-biasing it. When a pulse appears at base 1 of Q1, only the gate of SCR1 receives the pulse and only

**Parts List (Fig. 3-7)**

R1, R2: 500K linear potentiometers
R3, R4: 750K, 1/2 watt
R5: 100K, 1/2 watt
R6, R7: 1K, 1/2 watt
R8: 270 ohms, 1/2 watt
R9: 4.7K, 1/2 watt
R10: 250 ohms, 5 watts
C1: 0.47 mfd, 50 volts
C2, C3: 0.22 mfd, 50 volts
C4: 4 mfd, 50 volts, non-polarized electrolytic
Q1: GE 2N2646 UJT
SCR1, SCR2: GE C106F
D1, D2, D3, D4, D5: GE A13F
Load: GE 50C, 1.4 amp lamp

SCR1 is turned on. Capacitor C4 charges through load L1, with the positive side connected to the anode of SCR2 and the end which is nearly at ground potential connected to the anode of SCR1. With SCR1 on, the bias on D5 is removed, and the junction of diodes D1 and D2 is clamped to nearly 1 volt because D2 is now forward biased.

Capacitor C1 now starts charging through resistors R1 and R3 and diode D3. At the end of the time delay (which is adjusted by setting potentiometer R1) the unijunction produces another pulse. This pulse turns SCR2 on. When SCR2 turns on, it is the same as if capacitor C4 were connected across SCR1; in other words, it is momentarily reverse-biased. The momentary reversal of anode potential turns SCR1 off. With SCR2 on, diode D4 is forward-biased; therefore, capacitor C1 starts charging through R2, R4, and D1. With this arrangement, the off time (SCR2 off) is determined by the setting of R1 and the on time is determined by the setting of R2.

SCR2 should be selected so that the maximum load current is within its rating. Since SCR1 is used only for commutating SCR2, it can be smaller in rating than SCR2. Notice that the more current through the load, the larger the value of C4 would have to be. The minimum value of C4 can be determined from:

$$C4 \geq \frac{1.5 t_{off} I}{E}$$

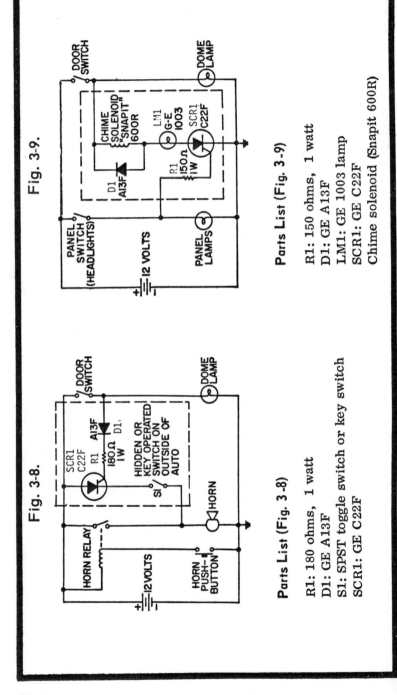

Fig. 3-9.

**Parts List (Fig. 3-9)**

R1: 150 ohms, 1 watt
D1: GE A13F
LM1: GE 1003 lamp
SCR1: GE C22F
Chime solenoid (Snapit 600R)

Fig. 3-8.

**Parts List (Fig. 3-8)**

R1: 180 ohms, 1 watt
D1: GE A13F
S1: SPST toggle switch or key switch
SCR1: GE C22F

where C4 is in mfd; $t_{off}$ is the turn-off time of the SCR in microseconds; I is the maximum load current (including possible overloads) in amperes at the time of commutation; E is the minimum DC supply voltage.

If the anode of SCR1 has a lamp in its anode as a load instead of resistor R10, the circuit as shown would not operate properly because when SCR1 is on, the trigger pulse is coupled to both gates; SCR1 would not have enough time to turn off because of the short time constant involved. With the component values shown, however, is not objectionable to trigger SCR1 and SCR2 at the same time because the time constant R10-C4 is much longer than the trigger pulse width so that SCR1 remains reverse-biased after the end of the trigger pulse long enough to assure commutation of SCR1. To be able to drive equal loads in the anodes of the SCRs, SCR1's gate (Fig. 3-7) needs a bias similar to SCR2's gate bias. If this is the case, some additional starting means has to be incorporated into the circuit.

## AUTOMOBILE BURGLAR ALARM

Here is an interesting electronic circuit that any automobile owner would like to have in his car. It is a circuit that turns your car horn into a burglar alarm (see Fig. 3-8).

If the car door is opened without first turning off switch S1 (it can be a hidden toggle switch or a key switch located outside the car) the horn blows. It will continue to sound until S1 is turned off.

The portion of the circuit outside the dashed lines in Fig. 3-8 is already in the car. The components inside the dashed line are the only ones that must be added to the existing wiring.

## CAR LIGHTS-ON ALERT

Here's a simple SCR circuit you can add to your car. It is a battery saver that protects you against dead batteries resulting from accidentally leaving your headlights on when you get out of the car. With this circuit installed (Fig. 3-9) if you have your headlights on and open the car door, a chime rings to remind you to turn out your lights.

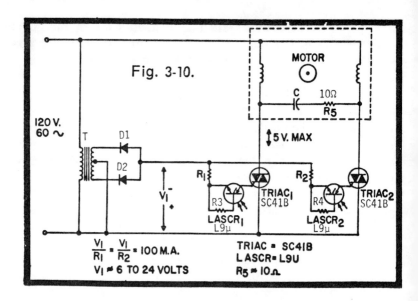

Fig. 3-10.

## Parts List (Fig. 3-10)

R1, R2: Resistance selected so that voltage V1 divided by resistance R1 or R2 equals 100 ma

R3, R4: Feedback resistance determined by the LASCR used

R5: 10 ohms

LASCR1, LASCR2: GE L9U

Triac 1, Triac 2: GE SC41B

T: Filament transformer: Primary, 120v AC; Secondary, 6-24v AC as desired.

## FOUR MOTOR-SPEED CONTROL CIRCUITS

First we have a reversing induction motor drive (see Fig. 3-10). In the circuit in Fig. 3-10 light-activated SCRs (LASCR) control the direction of rotation of a split-capacitor motor through two Triacs. Although this is basically an on-off control, it is used in proportional control systems where the motor drives some proportional control device.

Next, there is a speed control intended for use with shunt-wound DC motors (see Fig. 3-11). The phase control of this SCR is derived from the voltage across the SCR. Therefore, it effectively senses motor speed and loading conditions. This provides a form of feedback control for automatic motor speed regulation. The control circuit is reset at the end of each half cycle by resistor R1 and diode D1.

**Parts List (Fig. 3-11)**

R1: 1K
R2: 250K linear potentiometer
C1: 0.1 mfd
D1, D2: 1N1694
D3: A45B
D4, D5, D6, D7: A40B (4) bridge rectifier
SCR1: GE C22B
Triac 1: GE ST-2

The third circuit is a universal motor adjustable speed drive (Fig. 3-12). It is an extremely simple, yet sophisticated feedback control system that compares the voltage across the capacitor with the generated EMF in the armature to regulate motor speed. This circuit is widely used in power tools and appliances such as drills, saws, sewing machines, mixers, etc. To avoid reconnection of the motor, the series field may also be placed in the armature circuit. This permits plug-in connection to most universal motor devices.

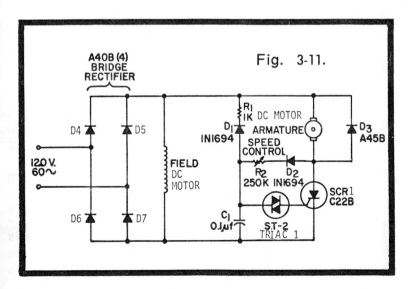

Fig. 3-11.

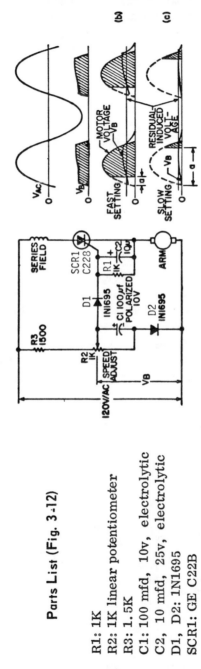

Fig. 3-12.

## Parts List (Fig. 3-12)

R1: 1K
R2: 1K linear potentiometer
R3: 1.5K
C1: 100 mfd, 10v, electrolytic
C2: 10 mfd, 25v, electrolytic
D1, D2: 1N1695
SCR1: GE C22B

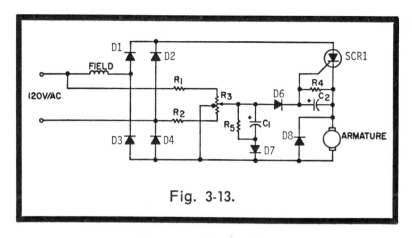

Fig. 3-13.

Parts List (Fig. 3-13)

R1, R2: 1.5K  
R3: 3K, center tapped  
R4: 1K  
R5: 470 ohms  
C1: 200 mfd, electrolytic  

C2: 5 mfd, electrolytic  
D1, D2, D3, D4: GE A40B  
D6, D7: 1N1694  
D8: GE A40B  
SCR1: GE C20B  

Motor: Series, reversible

The last circuit shown here (Fig. 3-13) is a reversible half-wave speed control. With a bridge rectifier feeding the SCR, armature current is always in the same direction. Field current, however, depends upon which polarity of the applied voltage is present at the time the SCR conducts. The direction of armature rotation is determined by causing conduction to occur in either the positive or negative half cycle of the supply voltage. Feedback control of motor speed is provided by comparing counter EMF with a capacitor voltage.

# SECTION IV

# Gadgets for the Home

ELECTRONIC BACHELOR LIGHT

Here's a great circuit to have going the next time you invite your girl friend over for an evening. Flip the switch and the controlled lamp slowly, slowly, slowly, slowly, gets dimmer, and dimmer, and dimmer, and dimmer. (See Fig. 4-1.)

The control voltage pedestal height is determined by the DC voltage across a large value capacitor (C1) through a Darlington emitter-follower circuit. Due to the high input-impedance of the Darlington connection, the capacitor charge and discharge times are very long, which produces very low turn-on and turn-off of the lamp load. At the maximum time setting, approximately 20 minutes is required to make the full transition from on to off condition.

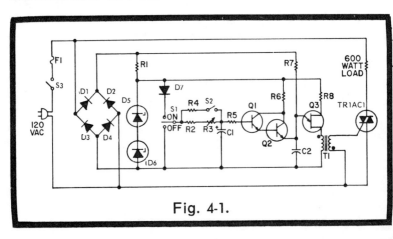

Fig. 4-1.

## PARTS LIST (FIG. 4-1)

R1: 3.3K, 2 watts
R2, R4: 4.7K, 1/2 watt
R3: 500K, 1-watt potentiometer
R5, R7: 1 megohm, 1/2 watt
R6: 2.2K, 1/2 watt
R8: 470 ohms, 1/2 watt
C1: 100 mfd, 15v electrolytic
C2: 0.1 mfd, 15v
D1, D2, D3, D4: GE 1N1693
D5, D6: GE Z4XL7.5 Zener diode
D7: GE 1N1692
Q1, Q2: GE 2N2712
Q3: GE 2N2647 unijunction transistor
Triac 1: GE SC41B
F: 3 amp fuse
S1: SPDT switch
S2, S3: SPST switch
T1: Pulse transformer, Sprague 35ZM923

## HIGH-GAIN LIMITED-RANGE LIGHT CONTROL

For applications in which the control of a load below the half-power point is not required, a rectifier can supply 1/2-cycle uncontrolled and an SCR can provide regulation by phase con-

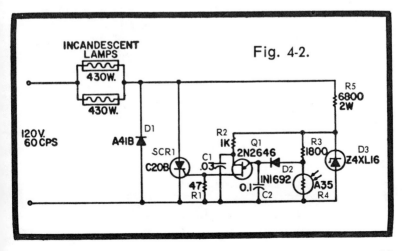

trol in the other half cycle. (See Fig. 4-2.) This is also a "ramp-and-pedestal" control system; the ramp amplitude is very small due to the logarithmic characteristic of the diode. The photocell is operated in a low-resistance condition in order to achieve fast response time and eliminate "hunting" when controlling the output of the lamp load.

## PARTS LIST (FIG. 4-2)

R1: 47 ohms
R2: 1K
R3: 1.8K
R4: Photoresistor, GE A35
R5: 6.8K, 2 watts
C1: 0.03 mfd
C2: 0.1 mfd
D1: GE A41B
D2: 1N1692
D3: Zener diode, GE Z4XL16
SCR1: GE C20B

ISOLATED LOW-VOLTAGE REMOTE CONTROL

The circuit in Fig. 4-3 provides convenient remote control via a low-voltage (6.3 volts AC) circuit. When remote switch S1 is open, transformer T1 blocks trigger current to Triac 1. With S1 closed, T1 saturates and triggers Triac 1. Typical applications include remote control of appliances, signaling systems, outdoor lighting, and remote controls for heating, ventilating, pumping, refrigeration, etc.

## PARTS LIST (FIG. 4-3)

R1: 50-ohm linear potentiometer
F: 5 amp, 3AG, fuse
J1: AC female receptacle (500 watts maximum)
S1: SPST switch
T1: Filament transformer: Primary, 120v; Secondary, 6.3v
Triac 1: GE SC41B

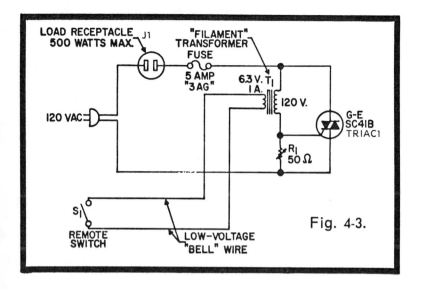

Fig. 4-3.

## FREQUENCY-SELECTIVE AC AMPLITUDE CONTROL CIRCUIT

Since the emitter-follower control of pedestal height may be regulated by a DC signal, power to a load can be controlled from any signal that can be converted to DC. For example, the amplitude of a tone signal can be used to control load power. The circuit in Fig. 4-4 can be used for remote control work, program control from a tape recorder, or for audio modulation by noise or by music with wide-band filters. The circuit may also be used with a tachometer for motor-speed control. No parts list is included with this circuit because the values vary widely according to the application. A true experimenter will have little trouble determining the exact values he needs.

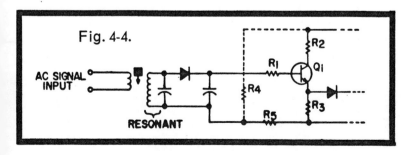

Fig. 4-4.

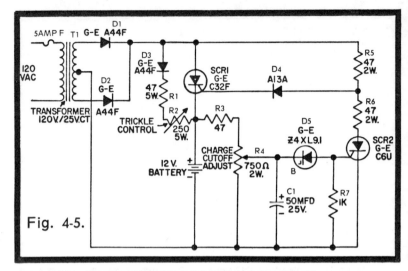

Fig. 4-5.

15-AMPERE BATTERY CHARGING REGULATOR

In battery-charging systems the circuit in Fig. 4-5 automatically cuts off heavy charging when the battery voltage reaches a preset level. The circuit allows charging to begin again when the voltage falls below a preset level. This circuit is particularly useful for portable tools, appliances, house and travel trailers, outboard marine use, emergency power, etc.

## PARTS LIST (FIG. 4-5)

R1: 47 ohms, 5 watts
R2: 250 ohm, 5-watt linear potentiometer
R3: 47 ohms
R4: 750-ohm, 2-watt linear potentiometer
R5, R6: 47 ohms, 2 watts
R7: 1K
C1: 50 mfd, 25v
D1, D2, D3: GE A44F
D4: GE A13A
D5: Zener diode GEZ4XL9.1
SCR1: GE C32F
SCR2: GE C6U
F: 5 amp fuse
T1: Filament transformer: primary, 120v; secondary, 25v centertapped

3-POSITION POWER CONTROL

Here are two simple circuits that are ideal for controlling loads such as heaters, ovens, lamps and universal motors. Both circuits provide three control positions: one off position, one low-speed or heat position, and one full speed or heat position. The circuit in Fig. 4-6 uses only a simple power silicon diode rectifier, and the circuit in Fig. 4-7 has an added Triac. This makes it ideal for cyclic control, low-maintenance applications, and wherever low noise or power gain is desirable.

## PARTS LIST (FIG. 4-6)

D1: GE A44B
S1: 3-position power switch
The load can be any 120-volt resistive load up to a maximum power rating of 5000 watts.

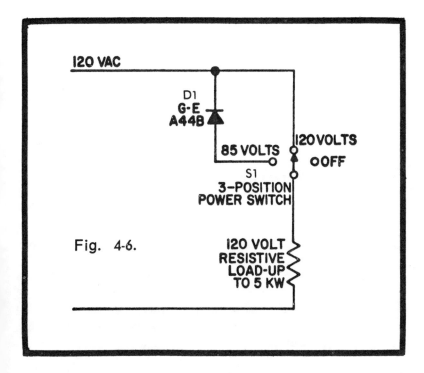

Fig. 4-6.

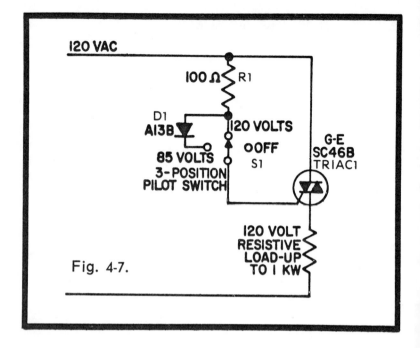

Fig. 4-7.

**PARTS LIST (FIG. 4-7)**

R1: 100 ohms
D1: GE A13B
S1: 3-position pilot switch
Triac 1: GE SC46B

The load can be any 120-volt resistive load up to a maximum power rating of 1000 watts.

HIGH-POWER BATTERY-OPERATED FLASHER WITH
PHOTOELECTRIC CONTROL

The circuit in Fig. 4-8 has a 36- to 40-watt power output, a variable flash rate of up to a maximum of 60 flashes per minute, independent control of on and off times, and a photoelectric daylight control to shut the flasher off during the day to increase battery life. The unit uses a very low cost GE C15FX321 SCR rated at 4.7 amperes average current at 60 degrees C stud temperature, 50 volts prv/vbo, maximum trigger current at 25 degrees C is 40 ma.

## PARTS LIST (FIG. 4-8)

R1: 1-megohm linear potentiometer
R2: 500K linear potentiometer
R3, R5, R6: 47K
R4: 47 ohms
R7: 100 ohms
R8: 1K
R9: 680 ohms
R10: 220 ohms
R11: Photoresistor, GE A33
C1: 20 mfd, 400v electrolytic
C2: 0.22 mfd, 150 volts
C3: 10 mfd, 25v electrolytic
D1, D2, D3: 1N1692
Q1: 2N2646 unijunction
SCR1, SCR2: C15FX321
LM1, LM2: No. 1034 lamps
Batt: 12 volts

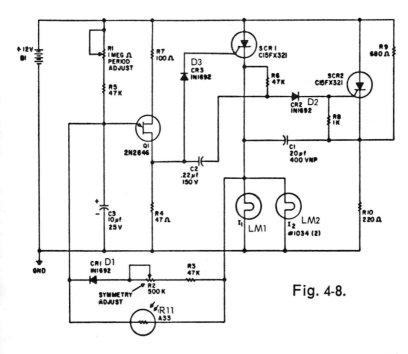

Fig. 4-8.

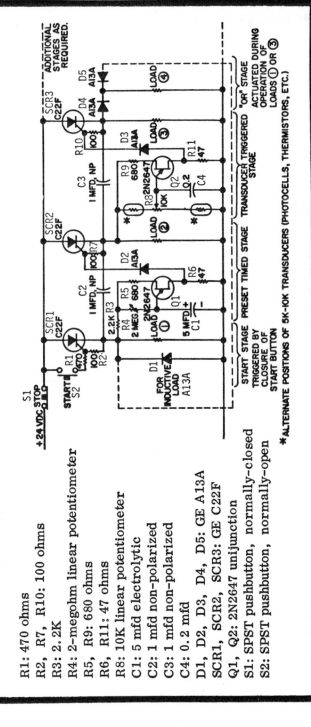

Fig. 4-9.

PARTS LIST (FIG. 4-9)

R1: 470 ohms
R2, R7, R10: 100 ohms
R3: 2.2K
R4: 2-megohm linear potentiometer
R5, R9: 680 ohms
R6, R11: 47 ohms
R8: 10K linear potentiometer
C1: 5 mfd electrolytic
C2: 1 mfd non-polarized
C3: 1 mfd non-polarized
C4: 0.2 mfd
D1, D2, D3, D4, D5: GE A13A
SCR1, SCR2, SCR3: GE C22F
Q1, Q2: 2N2647 unijunction
S1: SPST pushbutton, normally-closed
S2: SPST pushbutton, normally-open

SEQUENCE PROGRAMMER AND POWER DRIVER

A programmer circuit adaptable to timed or transducer-triggered intervals is shown in Fig. 4-9. Ideal for industrial process control, traffic controls, appliances, reproduction and photographic processes, vending machines, etc. With the indicated 1-mfd commutating capacitors, loads up to 1 ampere can be handled. Typical loads for this circuit include heaters, motors, solenoids, valves, relays and lamps.

35-WATT AUDIO AMPLIFIER USES PLASTIC TRANSISTORS

Complementary output transistors are a substantial aid in designing high-performance audio amplifiers, and with complementary silicon devices in plastic packages the component cost of high-quality audio amplifiers is substantially reduced.

The circuit shown in Fig. 4-10 will deliver 35 watts of music power into an 8-ohm load with low distortion and flat frequency response.

The 8-transistor amplifier is functionally divided into three sections: input circuitry, driver, and output. The input device is half of a dual transistor (two carefully matched transistors in a single package). The other half of the dual transistor provides feedback from the amplifier output. These two devices act as a differential amplifier. Driver transistor Q3 senses the collector current of input transistor Q1.

The 100-ohm collector resistor is used to limit the power dissipation in transistor Q3 during load fault conditions. Transistor Q4 is an active current source for Q3. The use of an active current source eliminates the large electrolytic capacitor usually needed to provide AC current during peak signal excursions. Compared to circuits using a capacitor, the active current source results in less distortion at low frequencies and provides a more symmetrical output.

The output circuitry is organized in two parts. Transistors Q5 and Q7 provide the positive output excursions, while Q6 and Q8 provide negative-going signals. This particular output configuration was selected as being the least sensitive to normal device parameter variations and offering the best linearity. Fig. 4-11 lists the amplifier's performance characteristics and identifies the Motorola semiconductors used in this circuit. The other component values appear on the schematic in Fig. 4-10.

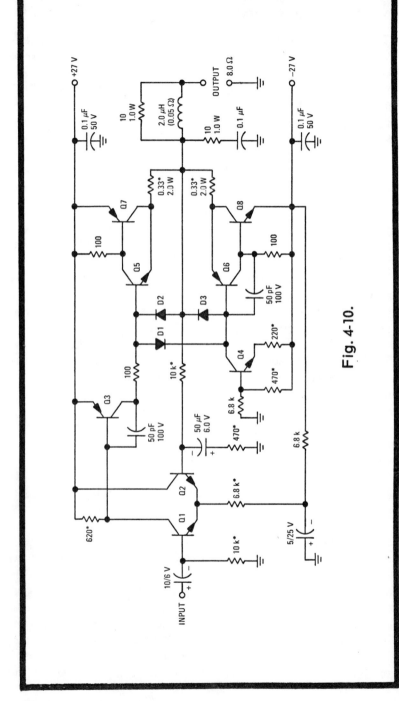

Fig. 4-10.

## 30-WATT 175-MHz POWER AMPLIFIER

The VHF power amplifier circuit in Fig. 4-12 uses PNP transistors to provide simplified construction in positive-ground systems  Until recently, RF power devices were available only as NPN units. Previously, if a system were operated at a positive ground, either the power amplifier chassis had to be ioslated from the rest of the equipment or performance suffered and the devices operated with greater than optimum emitter resistance.

The circuit shown in Fig. 4-12 provides 29.3 db of gain with only three transistors. (Partial specs of the amplifier are listed in Fig. 4-13.) The Motorola 2N5160 operates as a predriver for the 2N5161. The 2N5161, in turn, provides drive for the Motorola 2N5162 output transistor. Both the 2N5161 and 2N5162 are in stud-mount packages with the emitter connected to the stud. This is important for Class C

```
AMPLIFIER PERFORMANCE

Output Power:            35-watts music power into an 8-ohm load.  (RMS output
                         depends upon regulation of supply used.)

Sensitivity:             1v RMS into 10K for full rated output.

Frequency response:      Less than 3 db roll-off from 10 Hz to 100 kHz,
                         referenced to 1 kHz.

Total harmonic distortion: Less than 0.15% for any frequency and at any power
                           up to full rated output.

Intermodulation distortion: Less than 0.15% at any power up to full rated output
1 kHz and 7 kHz mixed at
4:1 ratio

Damping factor* (Ratio of   Over 150 at any frequency from 20 Hz to 20 kHz.
nominal speaker impedance   (Measured at 2w output.)
(8 ohms) to amplifier
output impedance)

                DEVICES                              LEGEND

Q1 and Q2 -- npn dual    Q6 -- pnp dirver   *±5% tolerance
transistor MD8001        MPSU55             All resistance values in ohms, 0.5 W
                                            ± 10%, unless otherwise noted.
Q3 -- pnp predirver      Q7 -- pnp Output   Note:  If parasitic oscillations are
MPSA55                   MJE2901            encountered, it may be necessary to
                                            add a 1.0 nF capacitor directly
Q4 -- npn Constant       Q8 -- npn Output   across the bases of Q7 and Q8.
current source           MJE2801
2N4410                                      Recommended minimum hea  sinking
                         D1 -- Dual
Q5 -- npn driver         stabistor MZ2361   Q5, Q6      θca = 40° C/W
MPSU05                                      Q7, Q8      θca =  3° C/W
                         D2 and D3 --
                         Stabistors
                         MZ2360
```

**Fig. 4-11.**

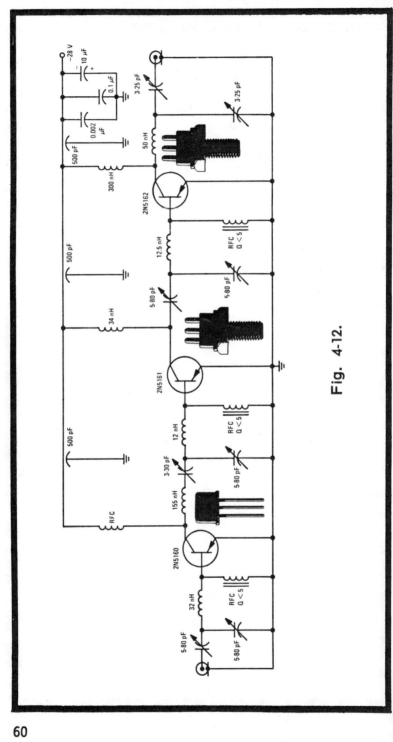

Fig. 4-12.

operation since a low-impedance emitter ground is needed to obtain maximum gain from each stage. With the stud-mount package, this ground is simply obtained by bolting the stud to a copper heat sink which, in turn, is fastened to the conductive chassis ground. This mounting technique also provides good heat transfer.

The three-stage amplifier delivers 30 watts of RF power into a 50-ohm load. Overall efficiency is quite high (50.5%) and all spurious outputs are at least 40 db below the 175-MHz level.

---

**Fig. 4-13.**

1) Input/output impedance: 50 ohms.
2) 2N5161 & 2N5162 should be heat-sink mounted.
3) 10W driving power is required.
4) The 3 500-pf capacitors are feedthrough types.

---

# SECTION V

# Trigac

The Trigac supplies systems designers, circuit engineers, and experimenters with an integrated circuit that facilitates the design of zero-crossing on-off control for a wide range of uses. Some applications include: motor on-off control, oven temperature control, motor protection, home-safety fire detector, etc.

The AC power control field has two major classifications for basic control systems—phase controls and zero-crossing switches (see Figs. 5-1 and 5-2). The phase control changes the average load power applied by varying the point at which the power switch is turned on. The top trace shows a small amount of power being delivered to the load; the lower trace shows almost full power being applied. Notice the abrupt change in waveshape at the thyristor's turn-on point.

One of the major problems with phase control systems is suppression of RFI (radio frequency interference) generated during the abrupt transition from the off to the on state. The amplitude of unwanted high-frequency energy generated is a function of the power being switched. Then the E-I product change determines the RFI intensity produced. Just before switching, the load current is zero, so the system's power will also be zero. This is represented by position A in Fig. 5-1. However, just after the thyristor has turned on, the system power condition is very high. In a 120-volt system, for instance, with a resistive load which dissipates 1200 watts average power, the instantaneous switched power may be as high as 1750 watts. This is represented at B in Fig. 5-1. Generally, this transient must by suppressed by filter networks. These filters have one or more components in the main load current conduction path that introduce significant power dissipation. Filter systems are generally costly, bulky, and may cause problems with reflected assymmetries in the power distribution system.

The zero-crossing switch has none of these problems. Its typical operation is shown in Fig. 5-2. Power is applied to the load by turning on the power switching thyristor when the line voltage (or current) crosses zero. With this approach, RFI is held to a minimum. (A direct inspection of the diagram shows that there will be a minimum energy change in the "before switch-on" to "after switch-on" transition.) The

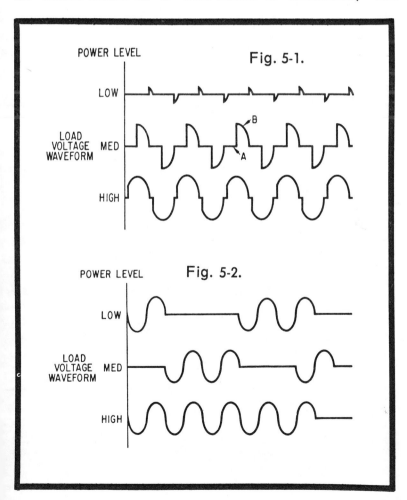

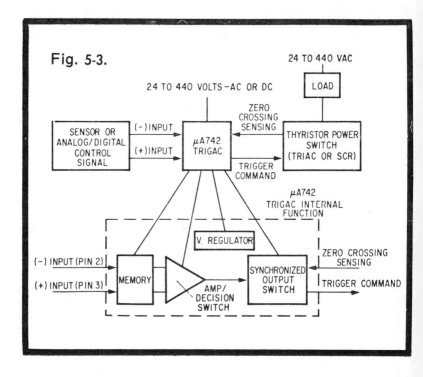

Fig. 5-3.

upper trace in Fig. 5-2 shows minimum power applied to the load; the lower trace, maximum power. Notice that the power input is increased by discrete steps of full cycles.

Designers or experimenters interested in developing controls with a minimum time investment will find enough data in the circuit application section that most closely resembles the need at hand and choosing the proper load switch (SCR or Triac) to match the system's power requirement. The Trigac's inherent flexibility and ease of application should enable circuits to be designed in this manner for functions ranging from the simple single-threshold coffee pot temperature control to the much more complex three-phase 24 KVA Y or delta-connected two-level threshold control complete with time proportioning. In all cases, wiring complication is kept to a minimum because the Trigac has its own internal power supply for operation with line voltages ranging from 24 to 440 volts (AC and DC). Also, it has internal voltage stabilization for compensating large variations in supply voltage. (For instance, when connected for 110-volt operation, the unit experiences practically no change in operation for line voltages between

85 and 135v AC.) Also, there is a built-in bias supply for external sensors which tolerate most commonly used analog sources. (Inputs such as unidirectional phototransistors or bidirectional photoresistors, temperature sensors, pressure sensors, etc., are acceptable.)

The Fairchild uA742 Trigac is a flexible integrated circuit interface between an analog sensor and the gate terminal of a power switching thyristor (SCR or Triac). Its location in a typical control system is shown in Fig. 5-3. The Trigac is designed to permit the maximum number of variations of this basic control configuration with the minimum number of external components. Its main function is to control the power applied to a load. The on-off decision is made in the following manner:

Whenever the positive input (pin 3) has a higher voltage than the negative input (pin 2), a current pulse will be delivered to the gate of the power switch at the following two load current zero crossings. This statement (illustrated in Fig. 5-4) contains enough information to accurately predict the control's response to all combinations of sensor input signals.

Control transfer characteristics can easily be described in electromechanical (relay) terms. For instance, in the presence of a slowly rising input (coil) signal, a normally-closed contact relay will reach a pickup voltage at which its armature will move from the unenergized condition to the energized condition. This causes its normally-closed contacts

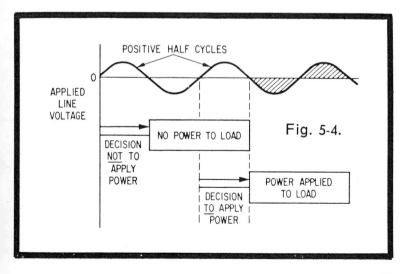

Fig. 5-4.

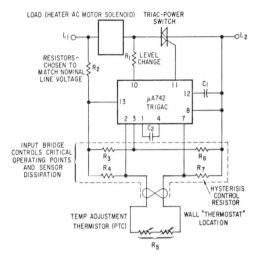

Fig. 5-5.

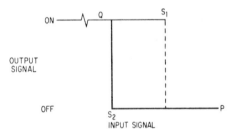

a. System transfer characteristics.

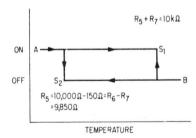

THESE VALUES ASSUME $R_3$, $R_4$, $R_6$ ARE 10kΩ

b. Critical operating point — two level control.

Fig. 5-6.

to open and results in the removal of power from the load. For higher values of applied signal at the relay coil, there will be no change in the load power. If the coil voltage is then lowered, a drop-out voltage is reached at some value below the pickup value. At this point, the armature returns to its unenergized position and load power is again applied.

The transfer characteristic of the relay's operation is shown in Fig. 5-6A. Switch S1 corresponds to the relay pick-up point while switch S2 describes the drop-out condition. It should also be noted that the load has only two possible conditions, on and off. This also establishes the shape of the basic zero-crossing hysteresis control characteristic.

The signals needed for the Trigac's on-off control decision are supplied by the input bridge network which contains resistors R3, R4, R5, R6, and R7 in Fig. 5-5. The R5 element is in a "remote" sensing location consisting of a PTC (positive temperature coefficient) thermistor temperature sensor in series with a temperature-adjust potentiometer. This branch of the input bridge would typically serve a "wall thermostat" type function. Notice that the circuit shows the sensor location as R5 (between Trigac terminals 2 and 7), although it could as well be in any of the input bridge's other three arms (represented by R3, R4 or R6). If an NTC (negative temperature coefficient) thermistor has been chosen, R6 would be the logical choice for the remote sensor position. If we assume that each arm of the input bridge is about 10K, then the critical operating points could be represented as in Fig. 5-6B.

TRANSFER CHARACTERISTICS

A typical system operating cycle (assuming that the controlled load in this instance varies the temperature in the vicinity of the input sensor) is as follows:

1. With the PTC thermistor resistance (R5) value low at low ambient temperatures, we enter the Fig. 5-6B diagram at Point A. The thyristor power switch is held in the on position by the Trigac.

2. As the thermistor is warmed by the heater element, the

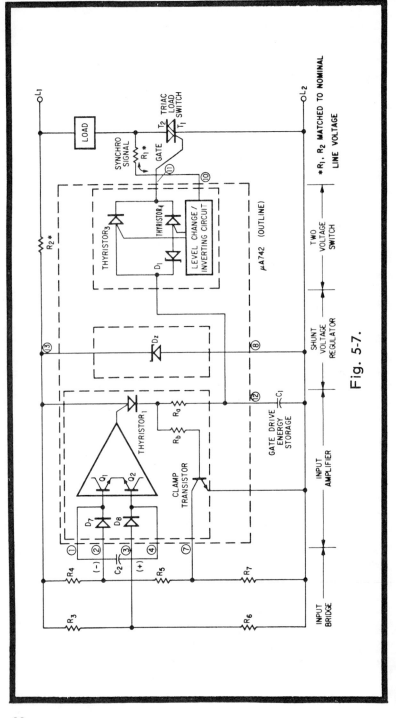

Fig. 5-7.

resistance of R5 is increased until the R5 plus R7 branch of the input bridge rises above 10K. At this point (S1), gate drive is removed from the power switching Triac, and the heater element (load) is turned off. The Trigac also shunts R7 with a current path approximately equivalent to a 150-ohm resistor.

3. After some thermal overshoot, the temperature at R5 decreases until the value of its branch of the input bridge is again at 10K (including the 150-ohm shunt at R7). We have now reached Point S2 in Fig. 5-6B and the Trigac again applies gate signal to the power switch.

4. After thermal undershoot, R5 will be rewarmed by the heater's output and the entire cycle will be repeated.

In Fig. 5-5 and the applications section to follow, each arm of the input bridge is shown to have a 10K resistance. Actually, there is a great deal of freedom in the choice of input components. The input bridge may be made up with values ranging from 4000 ohms to 40,000 ohms with little change in system performance.

THE uA742 CIRCUIT OUTLINE

The Trigac's principle of operation is explained in two parts:

1. A simplified description of the purposes of each functional block (Fig. 5-7).

2. Complete circuit description.

The circuit has two main sections: power supply, input amplifier, and a two-level synchornized output switch.

The power supply consists of a shunt zener regulator Dz. During time intervals in which the AC line is positive (L1 positive with respect to L2) this zener holds the maximum voltage at Vcc to within 21 volts of the system ground reference (L2). During the line supply's negative half cycle, the voltage at Vcc collapses and the portion of the Trigac to the left of the zener (the input amplifier, etc.) is allowed to "idle." We can say that during this, part of the circuit operates on a 50% duty cycle but that during each interval that the line applies over 21 volts (a period slightly less than 8.3 millisec), it operates as if it were being supplied from a normal DC source.

The input amplifier (and its associated functions) amplifies the input signal, makes the decision on whether or not power will be applied to the gate of the power switch by turning on an internal SCR whenever the positive input is larger than the negative input (thus defining the S1 value shown in Fig. 5-5), and sets the width of the hysteresis characteristic (the distance between S1 and S2 in Fig. 5-6) by shunting R7 in the input bridge. It retains a memory of the IC's operating condition from positive half cycle to positive half cycle.

The two input signals from the external bridge are fed through forward biased diodes D7 and D8 into the bases of transistors Q1 and Q2. Whenever the positive input exceeds the negative input by more than a slight offset voltage (typically about 3 mv), gate 2 current is extracted from thyristor 1. When this gate signal is present, thyristor 1 switches on, applying approximately 20 volts to resistors Ra and Rb. The resultant current through Ra charges capacitor C1, later used to supply gate drive for the power switch. At the same time, current through Rb turns on the clamp transistor so that the voltage drop across external resistor R7 is reduced to a single Vce (sat). This has the effect of driving the negative input further negative and widening the difference between the positive and negative inputs, which in turn furnished added gate drive for the switch.

The operation of the clamp (by shunting R7) also defines the change in input voltage level required to make the circuit return from the on to the off state.

The hysteresis transfer characteristic in Fig. 5-5 explains the need for D7, D8, and C2. For input signals between the two critical operating points, S1 and S2, the system may have either of the two possible output stages, on or off. For instance, if an input signal between S1 and S2 is applied for the first time, the system will assume the off state. However, if later changes in the input signal cause the system to turn on, then it should continue to retain the on condition for signals between S1 and S2 until the lower threshold point (S2) is reached. To do this, the circuit must have some form of memory. In conventional two-level circuits, such as a Schmitt trigger, a transistor latched on maintains a record from moment to moment of the system's previous state. Obviously, a continuous Vcc supply is needed to hold the latched transistor on for this type of memory. The Trigac, with its periodically interrupted Vcc, must resort to a different form of memory.

Memory of the control's condition is kept during negative line half cycles (when the circuit is idling) by energy storage in capacitor C2. This stored energy forces the differential amplifier (Q1 and Q2) to assume the previously held state at the beginning of each positive half cycle. Diodes D7 and D8 prevent C2's stored voltage from discharging into the input bridge during idle intervals. The charge on C2 is refreshed during each positive half cycle. C2 also has a secondary function: it shows the amplifier frequency response to help eliminate false system noise turn on. (Actually, system frequency response is set by the relatively slow 60-Hz line.)

To summarize, the input amplifier has the job of delivering energy to C1 whenever a gate signal is required for the external power switch. The decision on whether or not to supply this energy is made during each positive half cycle of the line. Once C1 has been charged, the two-level synchronized

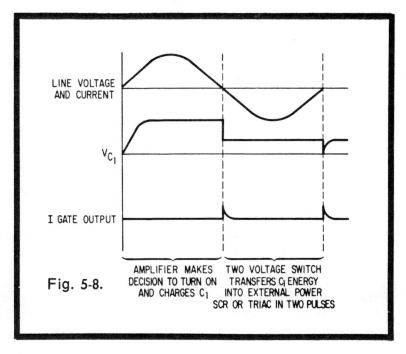

Fig. 5-8.

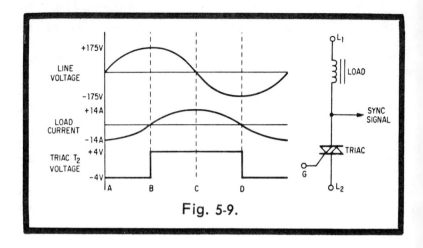

Fig. 5-9.

switch (covered next) will pass its stored energy to the gate of the following SCR or Triac with ten milliseconds.

The two-level synchronized output switch transfers the C1 charge into the external power switch gate during the two line-current zero crossings immediately following the particular positive cycle in which C1 was charged (see Fig. 5-8). To do this, two basic functions are required:

1. The energy must be parcelled out in two separate bursts; the first occurring when the load current crosses zero while traveling in the negative direction, and the second when the load current again crosses zero while traveling in the positive direction.

2. The amount of energy discharged from C1 during both pulses must be accurately controlled so that the external power thyristor receives an adequate gate signal at each zero crossing.

These features are implemented through the use of thyristor 3 and thyristor 4 in Fig. 5-7. The signal derived from the T2 terminal of the power Triac changes sign when the load current passes through zero.

A digression is in order here to explain the significance of using a signal derived from T2 to sense information about the load current. The line/load waveforms for a typical AC inductive load switching condition are shown in Fig. 5-9.

Beginning at the left edge of the illustration, assume that the thyristor in use is a Triac and that it is already in conduction in the negative direction. When the load current passes through zero, the Triac loses holding current and momentarily presents a high resistance to the series divider formed by the load and the thyristor. Since the load has a relatively low impedance, the remote thyristor terminal attempts to increase to line voltage. This produces a positive signal that is then coupled into the pulse generator via the synchronizing signal connection. The arrival of this signal causes the pulse generator to very rapidly deliver a pulse to the gate of the thyristor. The thyristor then resets into the on condition for another half cycle.

The benefit of using this arrangement is obvious if we assume that the phase lag of the load current varies. (This is a situation frequently encountered in the case of motor loads. As the motor-start winding is switched out, the phase lag of the motor can change by as much as 50 degrees.) If we assume that the position of the load current's zero crossing moves either forward or back in time, it is obvious that the synchronizing signal will also shift. (The thyristor waits to fall out of latch until its current passes through zero.) This will cause precisely the required change in the timing of the thyristor gate pulse to hold RFI generation to a minimum.

Returning to the discussion of the two-level output switch's operation: The synchronizing signal from T2 passes through R1 and a level change inverting circuit (not shown in our simplified diagram) which extracts current from the gate 2 terminal of thyristor 4 just after the current zero crossing. This causes thyristor 1 to turn on, discharging C1 via zener diode D1, the anode-cathode circuit for thyristor 4 and the gate (T1 circuit of the external Triac power switch). C1 is discharged until its voltage is too low to sustain current through thyristor 4 (which then falls out of latch because its anode current falls below $I_{ho}$). The level at which C1 stops discharging (8 volts) is held for the remaining negative half cycle of the load current (as shown in Fig. 5-9).

When the load current next passes through zero (now traveling in the positive direction) the level-change inverting circuit extracts current from gate 2 of thyristor 3. This unit (which does not have a series zener) discharges C1 into the thyristor gate via its anode-cathode circuit and the Triac's gate (T1

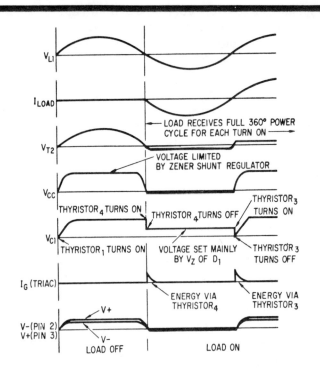

Fig. 5-10.

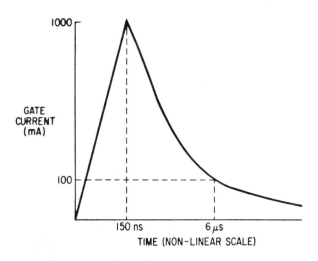

Fig. 5-11.

circuit). Thyristor 3 falls out of latch when C1's voltage has dropped to about 1 volt. The energy delivered into the gate of the power switch (Triac) has been controlled in both cases by the voltage change at C1.

At this point the line voltage becomes positive so that the input amplifier Vcc approaches 21 volts. If conditions at the Trigac's inputs (pins 2 and 3) dictate that thyristor 1 will again be turned on to charge C1, the entire cycle will be repeated for the following two load current zero crossings. The sequence of events during the uA742's operation generates the waveforms given in Fig. 5-10.

Now for a look at the circuit's signal output. The Trigac produces a nearly ideal gate drive for power switching thyristors. When attached to the recommended external components, its waveshape (shown in Fig. 5-11) has a rise time of about 150 nsec to a peak value of 1 ampere. The rate of decay is held so that there will be at least a 100 ma current available after 6 microseconds. If necessary, this interval may be stretched through alteration of the external components. This completes the simplified discussion of the Trigac's operation.


## TRIGAC OPERATION FROM AN AC SUPPLY

The schematic in Fig. 5-12 shows the connections for a Trigac control circuit operating directly from a single AC supply. It shows the necessary external components as well as the various sections within the Trigac. When operated directly from an AC line through a dropping resistor, zener diodes D2, D3, D4, and diodes D5, D6 in the power supply section provide a regulated supply of about 21 volts positive at pin 13 during the positive half cycles. During the negative half cycles, the isolation diode D1, holds this potential to about −0.7v.

The charge control section contains a conventional differential amplifier comprised of a matched pair of transistors Q1, Q2, and fed via a constant-current source (Q3). Q3 begins to conduct only after the supply voltage at pin 13 has exceeded about 14 volts or when diodes D3 through D6 conduct.

The inputs of the differential amplifier are connected to the

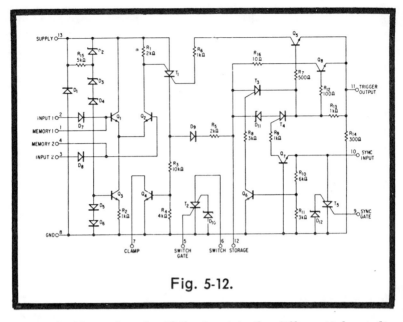

Fig. 5-12.

sensor bridge network. If the input to the differential amplifier is such that the voltage at pin 2 (negative input) is higher than that a pin 3 (positive input), Q1 will conduct and thus hold Q2 off. Let us call this the "inhibit state." When the input to the differential amplifier is positive (pin 3 at a higher potential than pin 2), Q2 will conduct and pull current out of the anode-gate of thyristor T1 in the charging network, causing T1 to start conducting. We will call this the "trigger state." As soon as T1 conducts, it will cause storage capacitor Cst to charge via T1, D9, and R5 to a voltage equal to the supply voltage of pin 13 minus the drop across T1 and D9; roughly 19.5 volts.

Transistors Q6, Q7, thyristor T5 and their associated components form the zero-crossing detector. They provide the necessary control to ensure that trigger output pulses supplied to the external circuit are delivered near the zero crossing of the load current to minimize the RFI generated. The trigger pulses are supplied through the pulse generator which receives its energy from the external storage capacitor Cst. The pulse generator is made of transistor Q8, thyristors T3, T4, and their associated components. The zero-crossing detector and the pulse generator work together and, therefore, their functions are explained simultaneously.

Early during positive half cycles, before the Triac anode voltage has reached about 7 volts, Q6 is forward-biased via resistors Rsyn and R10. When the anode voltage of the Triac exceeds the zener voltage of D12 (about 7 volts), T5 switches on and causes the sync input voltage at pin 10 to collapse to about 1 volt, thus turning Q6 off. This ensures that if T3 is to turn on, it will do so within the first seven or so volts of the positive half cycles. It will be recalled from the above explanation that Cst begins to charge only after diodes D3 through D6 have started conducting, or roughly when the supply voltage at pin 13 has reached about 14 volts.

Therefore, it is obvious that the charging and discharging of the storage capacitor occur at two distinct times. During the first positive half cycle, coinciding with, or immediately following the start of conduction of Q2, current will flow out of the anode gate of T3 within the first 7 volts of the positive half-cycle but will not turn T3 on due to the absence of voltage on storage capacitor Cst. A little later during the same positive half cycle when the supply voltage at pin 13 has reached about 14 volts, the storage capacitor will begin to charge. Its voltage will rise to about 19.5 volts during the remainder of the half cycle and will retain this value.

At the start of the following negative half cycle, Q7 begins to conduct and it causes T4 to turn on due to the current pulled out of the anode-gate of T4. T4 provides the base drive to Q5 and Q8. Thus, Q8 turns on and dumps some of the energy stored in Cst and provides a high-energy pulse to the gate of the Triac, causing the Triac to turn on near the beginning of the negative half cycle. As soon as the voltage across Cst falls to about 8 volts (because of D11), the current through T4 falls below its holding current level and thus T4 and Q8 are turned off. The storage capacitor, therefore, discharges from 19.5 volts to about 8 volts at the beginning of the negative half cycle.

Cst maintains this voltage for the remainder of the negative half cycle. At the start of the following positive half cycle, Q6 is forward-biased again, and T3 and Q8 conduct. Storage capacitor Cst now discharges to about 1 volt through R16 and Q8. Once the current through T3 falls below its holding current level, T3 turns off. The high-energy pulse so generated triggers the Triac on near the start of the positive half cycle. As soon as the voltage across the Triac collapses, the Q6

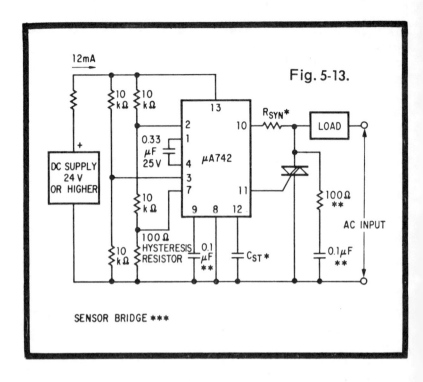

Fig. 5-13.

base drive and, consequently, the current out of the anode-gate of T3 is stopped. When Q5 comes out of saturation, Q2 makes T1 turn on again. The storage capacitor charges up and thus the cycle continues until the device reverts to the inhibit state.

OPERATION FROM A DC SUPPLY

The connection diagram for DC operation is in Fig. 5-13. There are some similarities as well as marked differences between DC and AC operations. In the DC operation mode, the constant supply voltage keeps the differential amplifier always in operation. Therefore, regardless of the instantaneous polarity of the AC line, storage capacitor $C_{st}$ starts charging as soon as T1 is triggered on by Q2. Let us now examine the transition from the inhibit state occurring during a positive half cycle of the AC supply.

At the beginning of the next negative half cycle, Q7 will be forward-biased and, just as it was in the AC operation mode, T4 will turn on, forward-biasing Q8 and Q5. Q8 produces an output trigger pulse at the beginning of this half cycle while Q5 pulls current out of the cathode gate of T1, causing it to turn off. Notice that in the AC mode of operation, T1 turns off at the end of each positive half cycle due to the natural reversal of the line voltage. In this mode, since the Trigac is powered through a DC supply and Q2 is assumed conducting, T1 turns on again as soon as Q5 turns off and recharges Cst back up to about 19.5 volts. With the next half cycle (positive), Q6 is forward-biased and, similarly, a trigger output pulse is delivered to the Triac; Q5 pulls current out of the cathode gate of T1, turning it off. This time Cst discharges to about 1 volt. However, when Q2 turns T1 on again, Cst recharges back up to about 19.5 volts and thus the cycle continues.

Now assume that the transition from the inhibit to the trigger state takes place during a negative half cycle of the AC line. At the beginning of the next half cycle, Q6 will be forward-biased and this time T3 will turn on first, forward-biasing Q8 and Q5. The Triac will, therefore, conduct initially at the start of a positive half cycle. The rest of the operation is similar to the description given above. When the input to the differential amplifier reverts back to the inhibit state, the Trigac will stop delivering output pulses. The Triac will then start blocking, always beginning with a negative half cycle.

## HYSTERESIS AND TIME PROPORTIONING OPERATION

So far, the function of hysteresis transistor Q4 and memory capacitor Cmem have been deliberately omitted. However, as can be seen in Fig. 5-12, every time T1 is turned on, Q4 saturates due to the bias it receives through R3. In the meantime, Cmem charges, according to the input conditions at pins 2 and 3. When Q4 saturates, it shunts hysteresis resistor Rhys and Q2 turns on harder, thus supplying positive feedback to the differential amplifier. The memory capacitor then adjusts its charge according to this new input and "remembers" it for the next cycle. The transfer characteristic of this mode of operation is shown in Fig. 5-14. Notice that if the

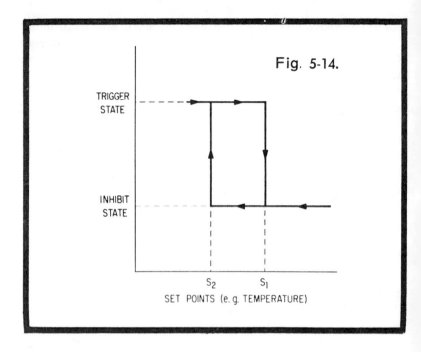

Fig. 5-14.

connection from pin 7 to the bridge input is omitted, Points S1 and S2 would coincide.

Proportional control, another feature available with the Trigac, is shown in Fig. 5-15. This function is achieved by using thyristor T2 to develop a ramp function which is superimposed upon the bridge input signal through the 200K resistor. Once the voltage on pin 6 reaches about 7 volts, T2 switches into conduction and lowers the input voltage to pin 3 and Q2 turns off. As the ramp voltage on pin 6 starts to increase again, the input to pin 3 also increases. The level on the ramp at which Q2 conducts determines how long the load remains energized.

APPLICATIONS OF THE TRIGAC

The following applications are a sampling of a few of the many Trigac uses. Each has an associated circuit board layout, parts list, and sufficient information for modeling the most common operating situations. The printed-circuit board layouts and parts list appear in the construction section at the end of this discussion.

Using printed-circuit boards to test Trigac circuits is encoutaged, particularly if voltages higher than 24v AC are to be used. When hooking up test systems with higher line voltages, the use of isolation transformers is recommended in the interest of safety.

ONE PHASE AC CIRCUITS

Application I: <u>110v AC Single Threshold Control</u>

This form of Trigac control circuit Fig. 5-16) requires the

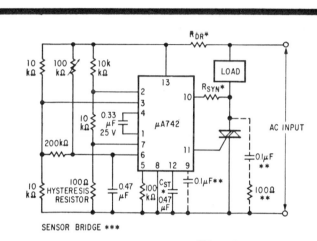

Fig. 5-15.

Recommended Values*

| AC Supply Voltage 60Hz Volts – RMS | $R_{DR}$ | $R_{SYN}$ | $C_{ST}$ |
|---|---|---|---|
| 24 | 1.0kΩ | 1.0kΩ | 0.47μF/25V |
| 110 | 10kΩ | 10kΩ | 0.47μF/25V |
| 220 | 22kΩ | 22kΩ | 0.47μF/25V |
| FOR SUPPLY VOLTAGE FREQUENCY OF 400Hz REDUCE $C_{ST}$ TO .047μF/25V | | | |

\*\* Necessary with inductive loads.

\*\*\* The sensor resistance will determine the values of the bridge resistors. For the values of $R_{DR}$ shown, the total current into the bridge should not exceed 5mA at 20V.

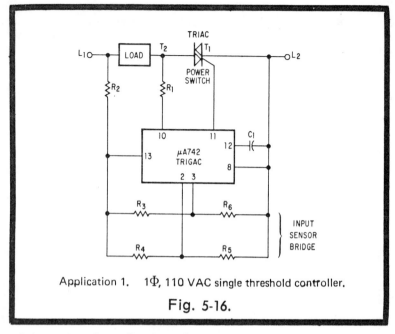

Application 1. 1Φ, 110 VAC single threshold controller.

Fig. 5-16.

minimum number of external components. Circuit operation is explained in the context of a simple low-cost temperature controller (a common application).

Assume that the input sensor is a PTC (positive temperature coefficient) thermistor with an operating point resistance of about 5000 ohms. Further assume that it is in series with a potentiometer (temperature adjust pot) set at a value near 5000 ohms. The value of R5 is then approximately 10,000 ohms (5000 ohms plus 5000 ohms). Then temperatures holding R5 below this value result in a positive input, a high or on condition. Temperatures holding R5 above this value provide for an off condition.

The circuit's operation is not limited to temperature control; the input sensor could as easily be a photodiode, photoresistor, pressure transducer, moisture sensor, water-level detector, or any other analog sensor with resistance values ranging from 200 ohms to 40K. For applications with bridge arm resistance values differing from 10K, the important factor to keep in mind is the relationship:

$$\frac{R3}{R6} \lessgtr \frac{R4}{R5}$$

Notice that one of these resistors would usually be a sensor in series with an adjustment potentiometer. Whenever the left-hand term of this equation is larger than the right, the external power switch will receive a zero-crossing gate drive from the Trigac. If the right-hand term is larger, then no gate drive is supplied to the power switch. In practice, there will be a very small undefined area when the two input signals (pin 2 and 3) are within a few millivolts of each other. In this case:

$$Vcc \left[ \frac{R3}{R3 + R6} - \frac{R4}{R4 + R5} \right] > 3mv$$

where $Vcc$ is 21v.

For most controls, this undefined region is too small to be significant.

<u>Application 2: 110v AC Dual Threshold Control (with Hysteresis; Fig. 5-17)</u>

The transfer characteristic, a square loop, with dual threshold "critical operating points" is described at the beginning

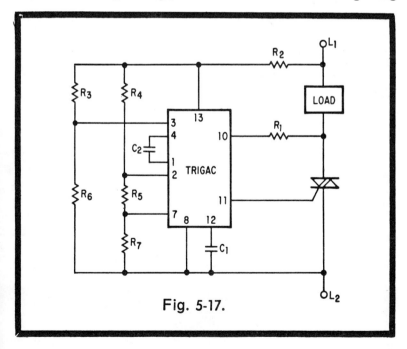

Fig. 5-17.

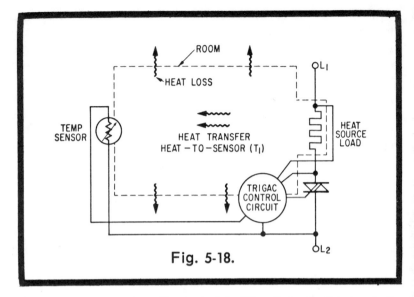

Fig. 5-18.

of this section and is illustrated in Fig. 5-6. In this case the input sensor values for both thresholds is set by the input bridge resistance values. The transition from the off to the on states is determined by the relative values of R3, R4, R5, R6, and R7, as discussed in Fig. 5-5.

1. Transition from off to on: In this situation, the clamp transistor of Fig. 5-7 is off and the input bridge is a simple 4-armed configuration with this relationship for bridge balance:

$$\frac{R3}{R6} \gtreqless \frac{R4}{R5 + R7} \quad \text{CONDITION FOR TURN ON}$$

When the bridge is unbalanced, so the left-hand term is larger than the right, the Trigac will supply power to the gate of the power switch (Triac or SCR). When turn on takes place, the input bridge will be unbalanced by an amount equal to the effect of replacing R7 with a Vce voltage drop.

2. Transition from on to off: Representing the effect of the clamp transistor in the circuit under consideration by inserting an equivalent resistor, R7* to replace R7:

$$R7^* = \frac{(R4 + R5)}{\frac{Vcc}{Vce\ (sat)} - 1}$$

This assumes that: input bias current can be ignored and that the circuit's offset voltage is zero; that the Trigac's internal Vcc is 21 volts; that the clamp saturation voltage is Vce (sat); and that the value of R7* is large compared to R7. The condition for turn off now becomes:

$$\frac{R3}{R6} \leq \frac{R4}{R5 + R7}$$

## Application 3: 110v AC Dual Threshold Control With Time Proportioning

The need for time proportioning (or for the ability to smoothly control the load power) can be appreciated if we consider the interaction between a control and the environment it controls. For instance, an example of a room temperature controller is seen in Fig. 5-18. There are a number of time related factors associated with this layout. If we list them in approximate order of importance:

T1—The primary path heat convection propogation time (period required for a step change in output at the heater element to change the temperature in the vicinity of the sensor). Typically more than 2 minutes.

T2—Response time for the temperature sensor to react to a step change in its ambient (15-45 seconds).

T3—Lumped summary of other effects, including: disturbances due to room air movement, boundary layer effects at the surface of the heater and the sensor secondary convection routes, changes in room heat loss due to external variations (outside temperature, etc.).

T1 and T2 are the dominant factors. A moment's reflection on the effect of the time delay between application of power to the heater and the responding resistance change in the thermistor sensor leads to the conclusion that the room's temperature must oscillate if power is applied in slowly cycling

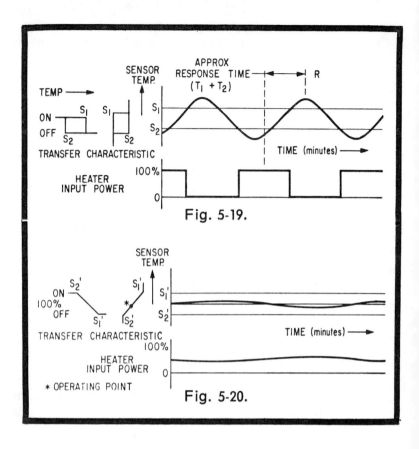

Fig. 5-19.

Fig. 5-20.

blocks of either full power or no power at all. This is illustrated in Fig. 5-19.

In many cases, tightened control of the hysteresis (S1-S2) will provide sufficiently accurate temperature control and time proportioning will not be needed. However, in systems which will not tolerate the overshoot-undershoot excursions of the dual threshold control, time proportioning is necessary. Here, the control should have a "single valued" response for each possible input temperature. Time proportioning is similar in concept to the continuously variable control in that the variation of the average power applied through a given input signal range (0 to 100%) is proportional to the sensor output.

If this response (Fig. 5-20) is compared to that in Fig. 5-19, the advantage of time proportioning is apparent. The time/temperature variation is smaller since the system stabilizes at a condition in which the average room power input exactly balances the heat loss. Therefore, virtually steady-state heat flow exists throughout the room. Circuits using phase control are capable of this feature but have the disadvantage of high RFI/EMI generation. Since zero-crossing control is by definition limited to turning the load supply on and off at the zero crossing of the load current, it must switch in whole cycles only. Therefore, the one possible method for varying the average load power input with zero-crossing control is to control the percentage of the total number of available whole line cycles which are applied to the load. One form of this technique is called time proportioning. Its operation is illustrated in Fig. 5-21. The top of this figure shows

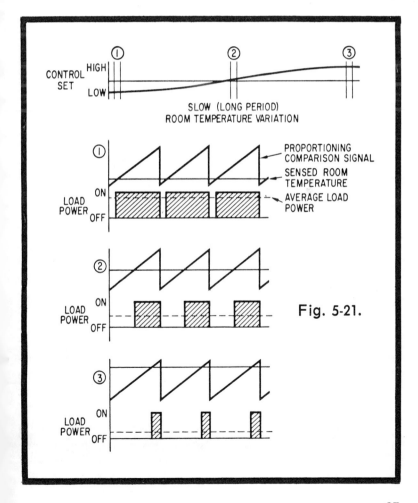

Fig. 5-21.

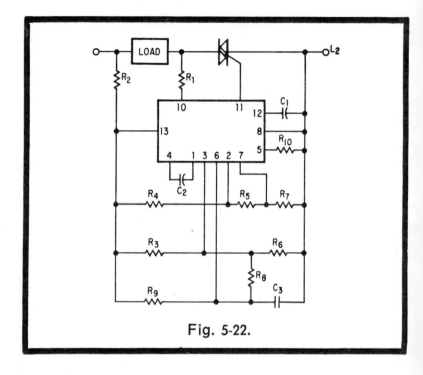

Fig. 5-22.

a room temperature control system in which the sensed temperature very slowly varies (possibly over a period of hours). Three representative conditions are selected:

Case 1. Low room temperature, the system calls for high heat.

Case 2. Medium room temperature, the system demand is for 50% heat.

Case 3. High room temperature, system demands low heat.

The first case is modeled by the signals shown in the second graph from the top in Fig. 5-21. Notice that the input temperature is represented by a straight line (the variation in this parameter is so slow that no slope is visible). The proportioning reference signal is represented by a sawtooth waveform. If the control system is arranged so that the load power is turned on whenever the sawtooth reference is higher than the sensed room temperature, then the "load power" cycling period will occur as shown. It should be kept in mind that the switching rate of the load is very fast relative to the room's response. Therefore, the power "bursts" shown on these diagrams are seen by the room as continuous power input (represented by the dashed line). In this control condition the average load power is high and the load is turned on for a greater proportion of the time.

The second case (shown in the third graph from the top) is similar to the first, with the exception that the temperature signal is higher and the control demands less power. In this situation, the proportioning reference input is higher than the measured temperature for only half of the time, creating conditions which hold the load power for an identical period. In Case 2, therefore, the room receives an average power equal to half of the power available when the heater is on continuously.

The third case shows conditions when the input temperature is high and the load power demand is low. Here, the proportioning reference exceeds the sensed temperature only near the peaks of the sawtooth and the load is switched on for only a small percentage (or proportion) of the time.

<u>The time proportioning mechanism</u>: The circuit in Fig. 5-22 produces the transfer characteristic given in Fig. 5-20 and shows how the power applied to the load is "time proportioned" to control the average power output. The operation is implemented by simply generating a sawtooth waveform (analogous to the "reference" waveform in Fig. 5-21), through the use of an RC charging circuit and a constant voltage breakover trigger built into the Trigac. The sawtooth is then applied to the positive input of the Trigac (pin 3) via a coupling resistor. The comparison room temperature signal fed into the Trigac's negative input (pin 2) in the manner used in Applications 1 and 2. Thus, electronic signals reproducing those shown in Fig. 5-21 are produced. The circuit is shown in Fig. 5-23.

The sawtooth is generated by the "relaxation oscillator" formed by R9, C3, and a 6.6-volt (approximately) fixed threshold thyristor. If we start with C9 discharged, the charge current through R9 causes C3's voltage to increase during each positive

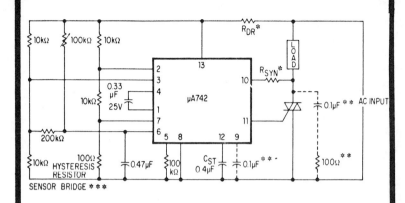

Fig. 5-23.

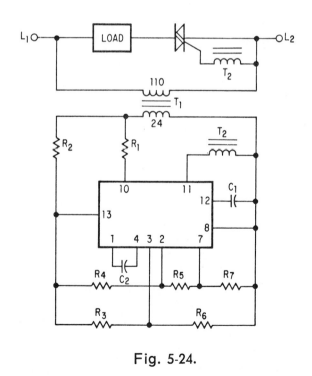

Fig. 5-24.

excursion of the line voltage (when the 21-volt Vcc is present). When the capacitor's voltage reaches the threshold 6.6 volts, the thyristor between Trigac pins 6 and 8 turns on and the capacitor is rapidly discharged to about 1 volt. The thyristor then loses latching current during a negative line half cycle and C3 is again permitted to begin charging.

The sawtooth thus generated across C3 is coupled to the pin 3 input of the Trigac by R8. The uA742's input amplifier section treats the sawtooth generator output as an additional input signal, which produces a tendency for the system to proportionally switch the load for sensor input signals varying between S1 and S2. This input signal's effect is set by the relationship of the resistance value of R8 to the other parts of the connected arms of the input bridge R3 and R6.

Since the time proportioning switch operates between two set points, V1 and V2 (the turn on and turn off voltages for the thyristor), and since the value of R8 is typically at least ten times the value of the standard bridge input resistors (R3 and R6), we can approximate the effect of the time proportioning switch's swing by saying that R8 is connected to ground at the end of the proportioning period (when the switch turns on). During the rest of the time, R8 is driven by C3's rising voltage ramp. If this is the case, then:

$$\Delta V \text{ PIN 3} = 21 \left[ \frac{R6}{R3 + R6} - \frac{R6//R8}{R3 + (R6//R8)} \right]$$

Substituting values for those in the construction section:

$$\Delta V \text{ PIN 3} = 21 \left[ \frac{5K}{5K + 5K} - \frac{\frac{(10K)(200K)}{210K}}{10K + 9.53K} \right]$$
$$= 21 (0.500 - 0.487)$$
$$= 0.237 \text{ V}$$

The circuit has a small resistor (R7) for the hysteresis function. This is used to provide the system with some noise immunity and should be used if any tendency to "jitter" is observed with slowly rising ramp inputs from the time-proportioning switch.

Notice also that a resistor (R10) has been placed between Trigac pin 5 (marked "switch gate") and the system ground. This resistor controls the switching sensitivity of the proportioning switch and should be selected for values between 10K

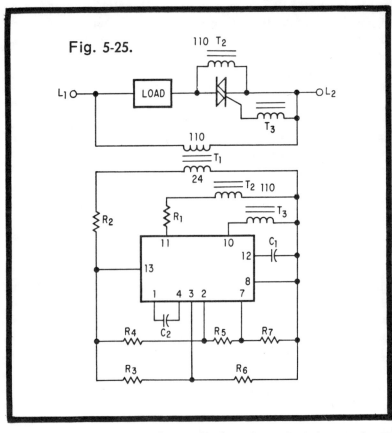

Fig. 5-25.

and 100K. In addition, for control systems in noisy environments (or when switching inductive loads) a small capacitor may be required to suppress noise pickup at the input bridge. If this is the case, a .01-mfd, 25v capacitor in parallel with R10 should eliminate the problem.

Application 4: Operation with Transformer Isolation Between The Control Circuit And The AC Line

The form of the Trigac's output gate pulses and the nature of the synchronizing signal make DC-isolated operation convenient. Two systems are shown: for circuits which have resistive loads (the synchronizing signal exactly in phase with the line voltage), and for circuits with inductive loads (synchronizing signal time shifted from the line voltage zero crossing).

The following circuits add to the flexibility of the four preceding applications. Each circuit has a printed-circuit board layout in conjunction with the construction section with letter-coded terminals that match directly with the PC board layouts for Applications 1, 2, and 3. Each accessory application has a separate part numbering system, so the parts list for each design should be consulted to avoid confusion.

Application 5: Output Pulse Amplifier

The Trigac produces an output pulse powerful enough for most currently manufactured Triac and SCR power switches. However, there may be situations in which very insensitive power switches or loads with extremely slow current rise times will require longer and larger gate current pulses. This circuit produces 2 ampere, 100 microsecond gate pulses

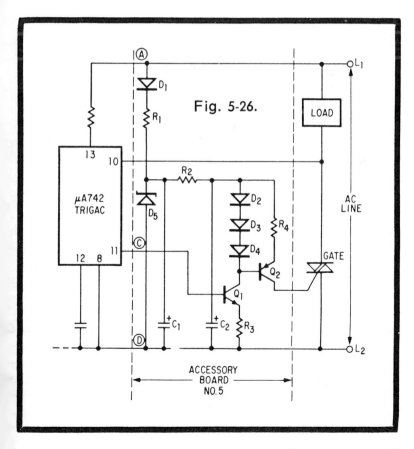

Fig. 5-26.

93

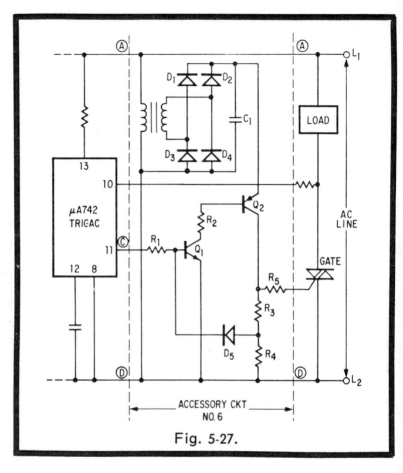

Fig. 5-27.

for a typical Triac and should be adequate for SCRs and Triacs with current ratings of over 400 amperes.

In Fig. 5-26, D1, R1, C1, and zener diode D5 form a 20-volt half-wave shunt regulated DC supply. (The circuit values are for a 110v AC supply line.) Energy stored in C2 is transferred via R4 and the collector-emitter circuit of Q2 into the attached Triac or SCR gate terminal. In a normal operating sequence, the output gate pulse from the Trigac is fed into the base of Q1. Q1 then turns on, forward biasing diodes D2, D3, and D4. This places a fixed voltage between the positive terminal of C2 and the base of Q2 and a relatively constant 1.8 volts between the higher voltage side of R4 and the base of Q2. If we subtract from this the 0.6 volt Veb of Q2, then the remaining 1.2 volts will be maintained across R4 and the emitter current will automatically be held at the value: 1.2/R4.

Since for reasonably high-gain transistors, Ie equals Ic, the combination D2, D3, D4, R4, and Q2 form a constant-current source which is switched on whenever Q1 is turned on by the Trigac's output pulse. Notice that the on period for this circuit can be controlled by varying the size of the Trigac's storage capacitor (C1).

Application 6: Output Pulse Amplifier (With Transformer)

This circuit (Fig. 5-27) produces amplified pulses of longer duration than those developed by Application 5. Also, it has lower power dissipation and lower cost for high-volume control systems. It produces output pulses which typically have a 2 ampere peak current with relatively linear decay to 1 ampere within 500 microseconds. The arrangement is basically simpler than that used for Application 5 but does not have the constant-current feature. Therefore, the output current is more dependent upon the power switch gate's terminal characteristics. For this reason, the circuit's output waveform should be checked with the particular SCR or Triac to be used.

Power for the circuit is supplied by the stepdown transformer

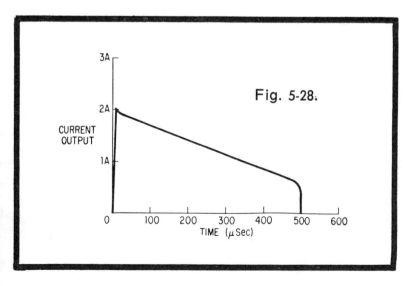

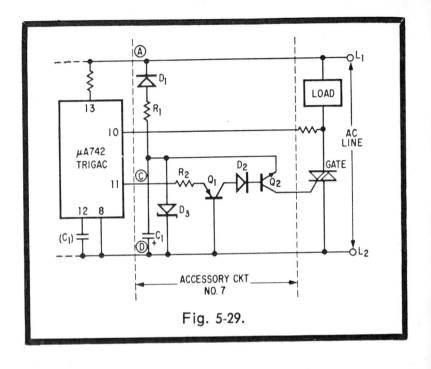

Fig. 5-29.

T1 (a common 6.3-volt filament transformer), the full-wave rectifier bridge formed by diodes D1, D2, D3, D4, and capacitor C1. When an output pulse is received from the uA742 via R1, transistor Q1 is turned on. Q1's collector current provides base drive (through R2) to PNP transistor Q2. Q2 is the output switch and it transfers energy from C1 into the gate of the external power thyristor. During the output pulse, the increased voltage at Q2's collector provides some added base drive for Q1 via the R3-R4 divider. This causes the Q1-Q2 transistor pair to latch on until C1 has been partially discharged. When the current supplied by C1 has fallen to a value below that required to hold Q1-Q2 in latch (about 1 ampere), the transistor pair turns off and C1 is recharged by the power supply.

Notice here that the energy contained in each output pulse is set by the voltage change C1 experiences during the pulse. The rate of discharge of C1 is set mainly by the Triac's input characteristics. Also, the Trigac serves only to initiate the output pulse for this arrangement and does not set the resultant Q2 on time. For this reason, the circuit is useful when it is desirable to lower the Trigac circuit's C1 (the charge capacitor), a good feature when driving insensitive gate load switching thyristors in 400-Hz systems.

Application 7: Output Pulse Inverter

Some Triacs, such as the logic Triac, require negative gate pulses for proper full-wave operation. Since the Trigac produces positive pulses, this circuit is included as an output pulse inverter for control systems using negative-gate thyristors.

The circuit's operation: D1, D3, R1, and C1 form a zener stabilized 20-volt half-wave supply for the inverter. Output pulses from the Trigac's pin 11 are fed through R2 into the emitter of Q1, permitting conduction in Q1's collector. This, in turn, supplies base drive to Q2 through D2. Q2 is then turned on, discharging the energy stored in C1 in the form of negative Triac gate current. The duration of this pulse is set by the Trigac's output pulse and may be varied by adjusting the value of R2 or the size of the Trigac circuit's C1. D2 prevents reverse breakdown of Q2's emitter-base junction.

Application 8: Period Extender for Time Proportioning

It is occasionally necessary to extend the period for the time proportioning (see Application 3) form of operation, to model external system time constants, for instance. The circuit given in this Application provides a proportioning time-base input for the Trigac with a period of up to 90 seconds.

The time base for the sawtooth waveform is generated by the RC charge circuit formed by R1 and C2 (Fig. 5-30). Charge from the Trigac's Vcc (21-volt) supply is supplied via forward biased D1 for this purpose. Tracing a standard operating cycle for generating the sawtooth:

1. Assume that C1 is charged and C2 discharged. Current through R1 caused C2's voltage to rise at the rate set by R1, C2.

2. This rising voltage is presented to the base of Q3's emit-

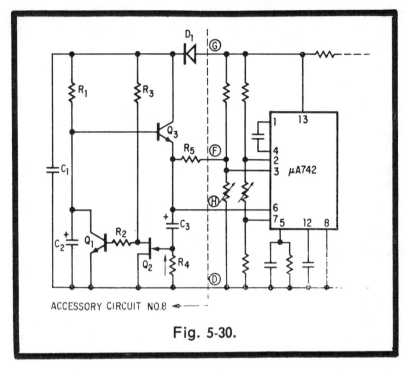

Fig. 5-30.

ter follower, which holds the positive terminal of C3 to within 0.6 volts of C2.

3. The rise of C2 is then reflected as an equal increase in C3's voltage, which is coupled by R5 to the positive input of the Trigac.

4. When C3's voltage has reached the breakover value of the Trigac's internal threshold thyristor, it is rapidly discharged via the path formed by pins 6 and 8 of the Trigac (the internal thyristor) and resistor R4. Discharge current flows through R4 in the direction shown by the arrow.

5. The charge-discharge fluctuation in the voltage of C3 generates a sawtooth waveform which functions in exactly the same manner as the shorter period sawtooth described in Application 3. R5 couples this into the positive input of the Trigac.

98

6. The remaining components have the function of resetting the charge in C2 to approximately zero during the C3 discharge. The method:

a. During the charging interval, current through R3 is shunted to ground by the drain-source circuit of Q2, a junction field-effect transistor. (For the purpose of this discussion, Q2 may be regarded as simply a resistor of about 125 ohms when zero or a positive voltage is applied to its gate. However, when a negative gate voltage of sufficient magnitude is applied, its drain-source terminals become essentially open-circuited.)

b. During the discharge interval: C3's discharge current through R4 produces a negative gate voltage for Q2, causing a sharp rise in the drain-source resistance. The current through R3 is then routed via R2 into the base of Q1. With Q1 turned on, C2 is reset via Q1's collector-emitter terminals.

Diode D1 and capacitor C1 serve to maintain the Vcc supply for this part of the circuit during the line voltage's negative cycles (when the Trigac's Vcc is not present).

Application 9: <u>Initial Cycle Delay</u>

A number of common AC loads have magnetic structures which are capable of being saturated during the first cycle after turn on. Loads of this type include:

Welding transformers

Large standard transformers

Variable autotransformers (Variacs)

Large motors

The first-cycle magnetic saturation results in a very sharp rise in load current just after the application of power. This phenomenon, often an unrecognized problem, can cause fail-

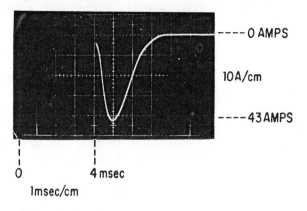

TURN ON CURRENT TRANSIENT WITH
UNMODIFIED ZERO CROSSING CONTROL

Fig. 5-31.

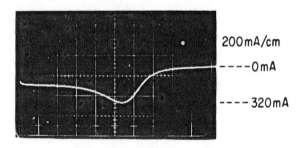

TURN ON TRANSIENT WITH FIRST CYCLE
TURN ON DELAYED FOR APPROXIMATELY 90°

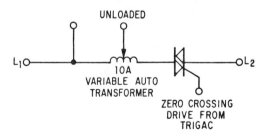

ure of the power-switching thyristor (Triac or SCR). For instance, the waveforms shown in Fig. 5-31 were taken with a standard 10 amp bench autotransformer. Notice that by delaying the start of the first cycle turn on of the Triac by about 4 msec, the surge transient has been reduced from 43 amperes to just about 0.3 amperes.

We may logically explain the use of initial-cycle delay in this manner: If we plot steady-state operating conditions for an inductive load, we get the results shown in Fig. 5-32. The three possible operating conditions are shown here. In the first plot, the steady-state (normal) condition shows that there is a positive current in the load at the extreme left of the diagram. This current crosses zero at approximately the peak of the negative half cycle and then reverses to reach a peak in the negative direction at the same time as the applied voltage's positive-going zero crossing. The initial positive current represents an energy storage condition in the load's core structure that "pre-biases" the load with a field which is reversed by the negative half cycle of the applied line voltage.

If we compare this with the center diagram (starting from the off condition) we see that no such "pre-bias" exists. In this case, the magnetic structure receives a full 8.3 msec of applied negative voltage. The current flow for the first quarter cycle of this operation is loosely analogous to time span A-B in the first graph. However, in this case the applied line voltage does not reverse at the end of a quarter cycle but continues with the same negative polarity. Therefore, after the first quarter cycle, further application of negative line voltage results in a rapid increase in load current, as shown. This causes an increase in magnetic flux until the core's saturation level is reached. The condition can be destructive to the Triac power switch, particularly if it coincides with a negative line transient.

The bottom plot demonstrates the effect of the initial cycle delay. A comparison of the top and bottom plot shows the similarity in the load current/flux condition. We have applied negative current to the load at approximately the same time that its current would have crossed zero when operating in steady-state conditions. The load receives only about 4 msec of applied negative voltage before the line reverses polarity. Numerous experiments have shown that this approach can reduce the turn-on transient to almost the same level as the

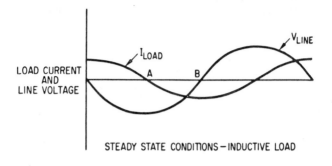

STEADY STATE CONDITIONS — INDUCTIVE LOAD

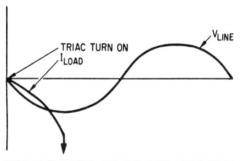

TURN ON TRANSIENT — WITHOUT INITIAL CYCLE DELAY

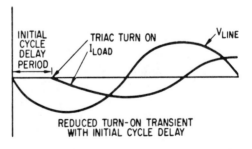

REDUCED TURN-ON TRANSIENT WITH INITIAL CYCLE DELAY

Fig. 5-32.

steady-state operating current. The RFI/EMI that would normally be generated by switching in midcycle is avoided because the flux-free load acts as its own inductive EMI filter during initial turn on.

The circuit in Fig. 5-33 blocks the first synchronizing signal to the Trigac's pin 10 terminal until the middle of the first negative line half cycle has been reached. If we assume that the Triac is off (there are no gate pulses from the Trigac), approximately full line voltage will be developed across its T1-T2 terminals. R4 and C1 will then form a phase-shift network which generates a 90-degree phase-shifted voltage across C1. This voltage is applied via R5 to the gate of Q2, a P-channel junction field-effect transistor. A bias network formed by R2 (connected to L1) and Q2's drain-source terminals drives the base of Q1 during the periods in which the Rds value of Q2 is high.

These conditions are illustrated in Fig. 5-34. In this circumstance, the circuit produces a synchronizing signal that is delayed by approximately 90 degrees from the line-voltage zero crossing. This condition continues until the Trigac's

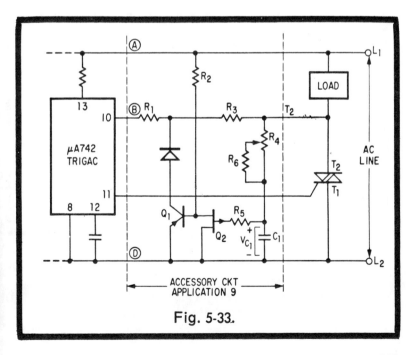

Fig. 5-33.

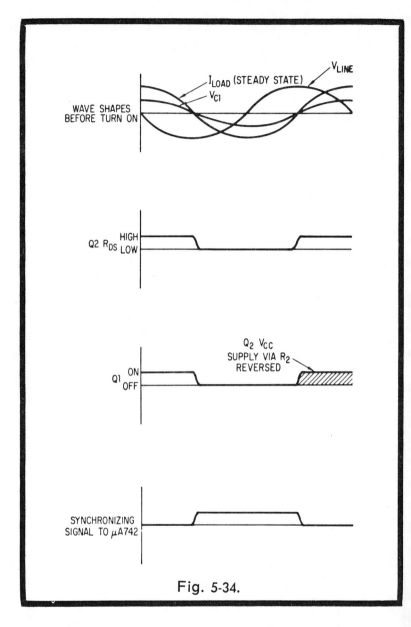

Fig. 5-34.

input amplifier calls for the application of gate drive to the Triac power switch.

When this occurs, the waveforms in Fig. 5-35 are generated. Here the circuit works as before until the drive to R4 is interrupted by the Triac's low saturation voltage (about 2 volts).

There is now insufficient drive for the R4, C1 phase-shift circuit to apply turn-off voltage to the gate of Q2. The FET then stays permanently on, holding Q1 off. With Q1 turned off, R1 and R3 act as the normal synchronizing drive resistor (which is given as R1 in the schematics for Applications 1, 2, 3, and 4).

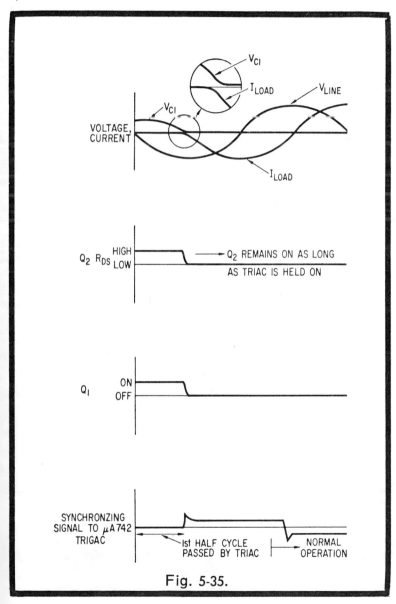

Fig. 5-35.

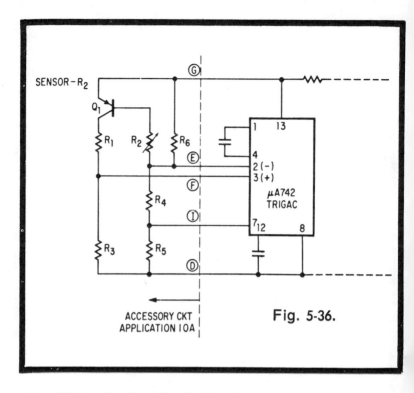

Fig. 5-36.

An additional note: The divider formed by R1 and R3 limits the Vce voltage applied to Q1. R1 also is a divider together with the Trigac's internal input resistors to prevent false application of the synchronizing signal. For optimum performance, the circuit's phase delay should be matched to the load. For this purpose, potentiometer R4 is provided.

Application 10: Sensor Failure Detection Circuit (Fail-Safe Operation)

In many control systems the failure (short or open circuit) of an input sensor can cause a dangerous condition. For instance, a heating control with an NTC (Negative Temperature Coefficient) sensor would interpret shorted thermistor leads as a very high sensed temperature and would interrupt the application of power to the load. This could be regarded as a "fail-safe" condition since furnace oven temperature (and the resultant fire or explosion hazard) is avoided. However, if the same NTC sensor fails in the open condition (due to lead wire breaks, etc.) the control system would respond as if to a low temperature, and power would be continuously applied to the load. In this case, an open sensor detector is required to protect the system against the resulting dangerous condition.

In general, both types of sensor (NTC and PTC) have one failure mode which is subject to interpretation by the control system in a dangerous manner. If we assume that the output thyristor off condition is safe, then the unsafe modes would be:

| Sensor Type | Unsafe Failure |
|---|---|
| NTC | Open circuit |
| PTC | Short circuit |

The two circuit modifications of the Trigac's input bridge (Figs. 5-36 and 5-37) are suggested to handle these conditions. Of course, other control situations (e.g., air conditioners or motor controls) will require a different combination of these two circuits. The important point is that these

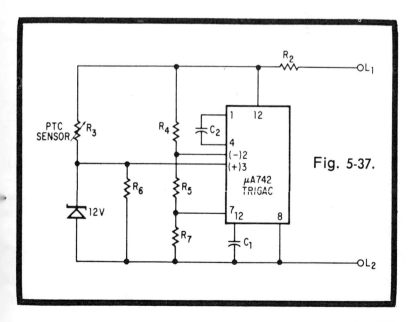

Fig. 5-37.

two techniques may be applied to detect either of the two possible dangerous conditions: short or open circuit.

During normal operation, Q1 is held in the saturated on condition by the current through R2. This provides enough base drive to apply Vcc (minus Q1's Vce(sat) to R1. R6 (about 50K) is too high to seriously affect the circuit's operation. If R2 (which includes the sensor) should open, then base drive for Q1 is interrupted and the transistor turns off. With Q1 off, there is no voltage supply for F1, and the positive input (pin 3) voltage falls. At the same time, R6 applies a positive voltage to the negative input. This fulfills the condition given previously, positive input lower than the negative input for the Trigac's off state, and insures that no gate drive will be delivered to the power switching thyristor, a "safe" condition.

The circuit for the shorted sensor detector is shown in Fig. 5-37. When the sensor arm of the input bridge is shorted, zener diode D1 limits the Trigac's Vcc voltage by shunting the IC's internal zener regulator. In this case, the Trigac's internal circuitry will prevent turn on of the input amplifier, and thereby prohibit charge transfer into the C1 storage capacitor.

It may be necessary to shift the values of R4 and R5 so that the positive input will not exceed 10 volts under normal operating conditions. No circuit layout is given for this in the construction section, since it merely involves adding a zener diode in parallel with one of the input bridge resistors for the printed-circuit boards given for Applications 1, 2, 3, and 4.

Application 11: Time-Delay "Relay" Circuit

There are many possible ways in which the Trigac's flexible input bridge may be used to generate a time-delay function. One of the most common utilizations would be the time-delay relay which is illustrated in Fig. 5-38. This circuit holds the Trigac in the off state for a controlled time after switch S1 has been opened.

If we assume that S1 is in the reset position, R5 holds the C2 voltage to within 1 volt of ground. This will hold Q2 in the off condition, effectively lowering the Trigac's positive input to near ground. At the same time the voltage divider formed by R1, R2, and R3 holds the negative input near 10 volts. The Trigac, therefore, will be held in the off condition.

When the switch is moved to the "Time" position, C2 is charged by current through R6. The emitter voltage of Q1 will correspondingly increase along with the C1 voltage until the positive input (pin 3) voltage is within several millivolts of the negative input. At this point, the Trigac switches into the on state and the gate C2 voltage will continue to rise until it stabilizes at some higher value. The circuit's C2 turn on voltage is set primarily by the resistor values in the bridge biasing the negative input (pin 2):

$$V\text{turn on} \simeq \hat{V_{cc}} \frac{R2 + R3}{R1 + R2 + R3} + V_{beQ1}$$

Where $\hat{V_{cc}} \simeq 21V$ and $V_{beQ1} = 0.5V$

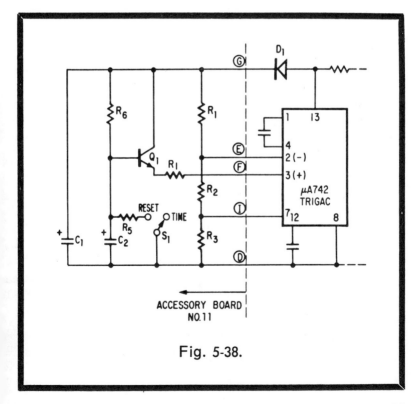

Fig. 5-38.

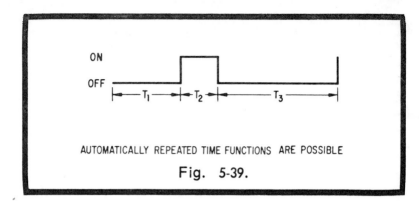

AUTOMATICALLY REPEATED TIME FUNCTIONS ARE POSSIBLE

**Fig. 5-39.**

Test results for the circuit values given in the construction section:

| R6 | C2 | Time Delay |
|---|---|---|
| 4.7 meg | 1.0 mfd | 5 seconds |
| 10 meg | 1.0 mfd | 10 seconds |
| 4.7 meg | 20 mfd | 150 seconds |

When S1 is returned to the "reset" position, C2 is discharged through R5. The value for R5 given in the construction section is small for the rapid reset of C2. However, if a time delay on the turn off function is desired, R5 may be increased. In this case, C2 would be discharged at the rate determined by the R5 x C2 product. The C2 voltage level for turn off would then be approximately.

$$V_{turn\ off} \simeq V_{cc} \frac{R2}{R1 + R3} + V_{ebQ1}$$

Where $V_{cc} \simeq 21V$ and $V_{beQ1} \simeq 0.5V$

D1 and C1 provide the timer circuit with a half-wave rectified DC supply. When the C2 voltage has reached this level, the Trigac will be held in the off condition until the switch is again moved to the "time" position.

There are three threshold detection methods for generating various simultaneous time-delay functions with the Trigac:

1. Normal input voltage level detection (as used in this application).

2. Use of the clamp transistor (pin 7) to generate a second turn off level by unbalancing the input bridge by a controlled amount.

3. Use of the Trigac's internal 6.6-volt proportioning switch (pin 6).

Combinations of these figures will produce control functions such as the one shown in Fig. 5-39. Also, automatically repeated time functions are possible.

## APPLICATION 1

| | | |
|---|---|---|
| $R_1, R_2$ | 1kΩ, 1W | (24 VAC Operation) |
| | 10kΩ, 2W | (110 VAC Operation) |
| | 22kΩ, 5W | (220 VAC Operation) |
| $R_3, R_4, R_5, R_6$ | 10Ω 1/4W | |
| $C_1$ | .47μF 25V Ceramic Capacitor | |

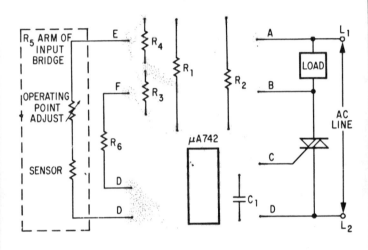

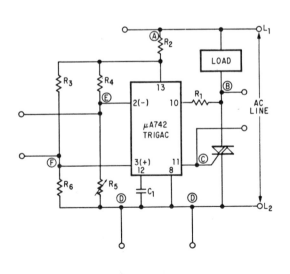

## Application 2 and 3.

## TRIGAC APPLICATION BOARD 2

| Part Number | Description |
|---|---|
| $R_1$, $R_2$ | 1kΩ 1W for 24VAC, 60Hz<br>10kΩ 2W for 110VAC, 60Hz<br>22kΩ 5W for 220VAC, 60Hz |
| $R_3$, $R_4$, $R_5$, $R_6$ | These four Resistances are in the input bridge and are nominally 10kΩ each. $R_5$ and $R_6$ are the variables in the bridge and either one may be used depending on the sensor used. |
| $R_7$ | The value of this Resistance determines the amount of Hysteresis in the System. (Min. 100Ω for 10kΩ Bridge). |

Proportioning Components

| | |
|---|---|
| $R_8$ | 200kΩ 1/4W |
| $R_9$ | 20kΩ 1/4W |
| $R_{10}$ | 39kΩ 1/4W |
| $C_3$ | 5µF 25V Electrolytic Capacitor |
| $C_4$ | .01µF 10V Ceramic disc. Capacitor |
| $C_1$ | .47µF 25V (60Hz Operation) Ceramic Capacitor Sprague 5C023474X025-0B3 (or equivalent) |
| | .047 (400Hz Operation) |
| $C_2$ | .33µF 10V Ceramic Disc. Capacitor Sprague HY-327 (or equivalent) |

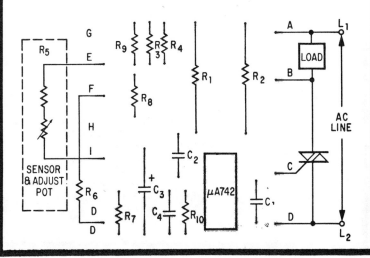

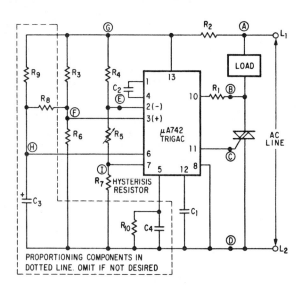

## APPLICATION 4

Use the PC board supplied for Applications 2 and 3, and mount isolating transformers off the board.

## ACCESSORY CIRCUIT APPLICATION 5

| Part Number | Description |
|---|---|
| $R_1$ | 2kΩ 5W Wirewound |
| $R_2$ | 270Ω 1/2W |
| $R_3$ | 10Ω 1/4W |
| $R_4$ | .5Ω 1/2W |
| $C_1$ | 20μF 50V Electrolytic Capacitor |
| $C_2$ | 25μF 25V Electrolytic Capacitor |
| $D_1$ | 1N 4004 (or equivalent) |
| $D_2, D_3, D_4$ | FAIRCHILD FD 222 (or equivalent) |
| $D_5$ | 1N 3796 (20V 1.5W Zener) |
| $Q_1$ | FAIRCHILD SE 6002 (or equivalent) |
| $Q_2$ | FAIRCHILD SE 8510 (or equivalent) |

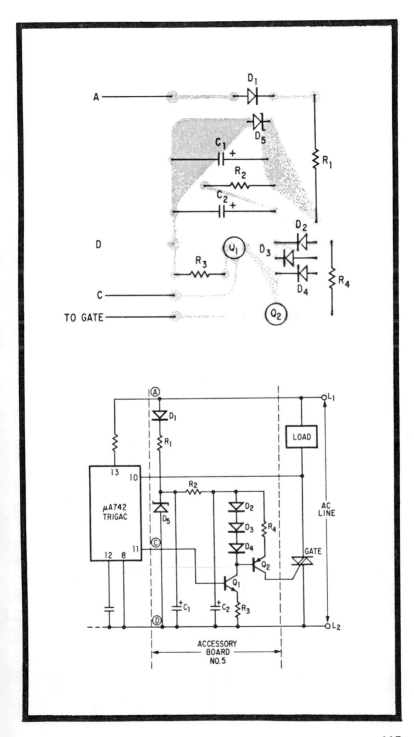

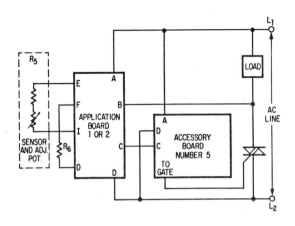

## ACCESSORY CIRCUIT APPLICATION 6

Part Number | Description
--- | ---
$R_1$ | 27Ω 1/4W 10%
$R_2$ | 20Ω 1/4W 10%
$R_3$ | 270Ω 1/4W 10%
$R_4$ | 120Ω 1/4W 10%
$R_5$ | 1Ω 1/2 10%
$C_1$ | 100μF 25V Electrolytic Capacitor
$D_1, D_2, D_3, D_4$ | Silicon diode In2069 (or equivalent)
$D_5$ | FAIRCHILD FD222 (or equivalent)
$Q_1$ | FAIRCHILD 2N3642 (or equivalent)
$Q_2$ | FAIRCHILD SE8510 (or equivalent)
$T_1$ | 110V/6.3V 1 Amp Filament Transformer

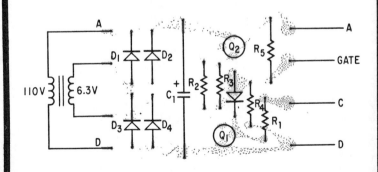

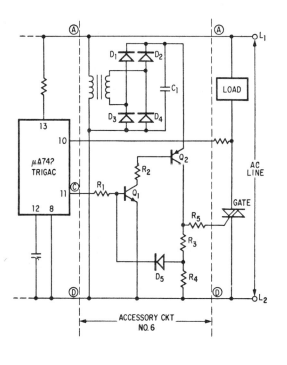

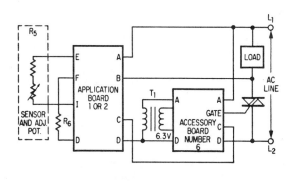

## ACCESSORY CIRCUIT APPLICATION 7

| Part Number | Description |
|---|---|
| $R_1$ | 5kΩ  5W |
| $R_2$ | 27Ω  1/4W  10% |
| $C_1$ | 5μF  25V Electrolytic Capacitor |
| $D_1$ | 1N2070 (or equivalent) |
| $D_2$ | FAIRCHILD FD222 (or equivalent) |
| $D_3$ | 1N4747 20V Zener Diode (1W) |
| $Q_1$ | 2N3638 (or equivalent) |
| $Q_2$ | 2N3642 (or equivalent) |

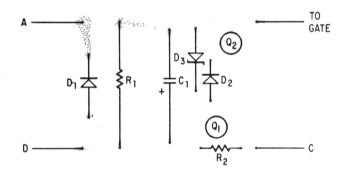

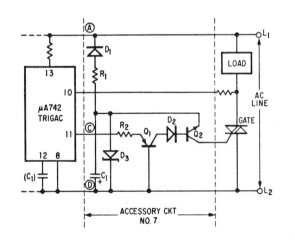

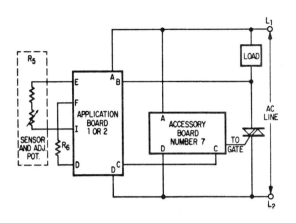

## ACCESSORY CIRCUIT APPLICATION 8

Part Number | Description
--- | ---
$R_1$ | 3.9MΩ 1/4W
$R_2$ | 10kΩ 1/4W
$R_3$ | 47kΩ 1/4W
$R_4$ | 270Ω 1/4W
$R_5$ | 330kΩ 1/4W
$C_1$ | .47µF 25V Ceramic Capacitor
$C_2$ | 25µF 25V Electrolytic Capacitor
$C_3$ | 1µF 25V Electrolytic Capacitor
$D_1$ | FAIRCHILD FD222 (or equivalent)
$Q_1$ | FAIRCHILD SE6002 (or equivalent)
$Q_2$ | 2N5163 (or equivalent)
$Q_3$ | FAIRCHILD SE4010 (or equivalent)

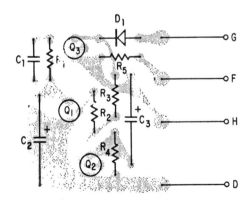

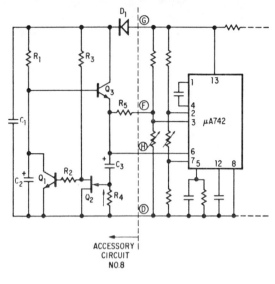

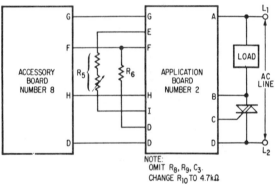

Application 9. Initial cycle delay diagram.

| Part Number | Description |
| --- | --- |
| $R_1$ | 1.5kΩ 1/2W |
| $R_2$ | 330kΩ 1/4W |
| $R_3$ | 8.2kΩ 2W |
| $R_4$ | 250k Trimpot, (IRC type) |
| $R_5$ | 4.7MΩ 1/4W |
| $R_6$ | 33kΩ 1/4W |
| $C_1$ | .025µF 200V Ceramic disc. capacitor |
| $D_1$ | FAIRCHILD FD222 (or equivalent) |
| $Q_1$ | 2N4250 |
| $Q_2$ | 2N4343 |

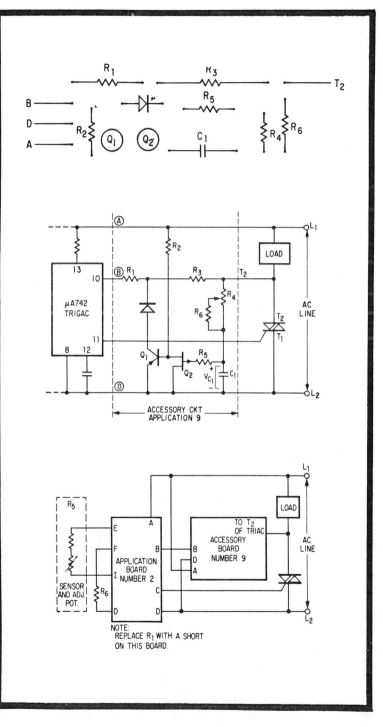

Application 10. Open sensor detector.

| Part Number | Description |
|---|---|
| $R_1$ | 10kΩ  1/4W |
| $R_2$ | Sensor (NTC) & adjust Pot. (Total of 10kΩ) |
| $R_3$ | 10kΩ  1/4W |
| $R_4$ | 10kΩ  1/4W |
| $R_5$ | 100Ω  (increase if more hysteresis is desired) |
| $R_6$ | 100kΩ  1/4W |
| $Q_1$ | 2N4250 (or equivalent) |

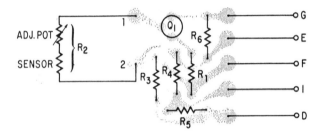

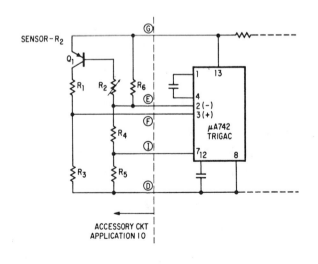

ACCESSORY CKT
APPLICATION 10

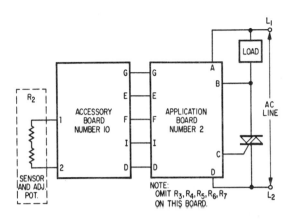

Application 11. Time delay relay.

| Part Number | Description |
|---|---|
| $R_1$ | 47kΩ 1/4W |
| $R_2$ | 100kΩ 1/4W |
| $R_3$ | 1kΩ 1/4W |
| $R_4$ | 47kΩ 1/4W |
| $R_5$ | 100Ω 1/4W |
| $R_6$ | Timing Resistor (See Text) |
| $C_1$ | 10μF 25V Electrolytic Capacitor |
| $C_2$ | Timing Capacitor (See Text) |
| $D_1$ | FAIRCHILD FD222 (or equivalent) |
| $Q_1$ | FAIRCHILD SE4010 (or equivalent) |
| $S_1$ | SPDT Switch |

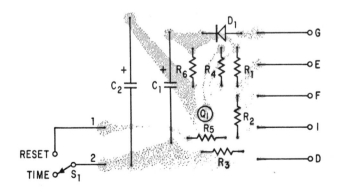

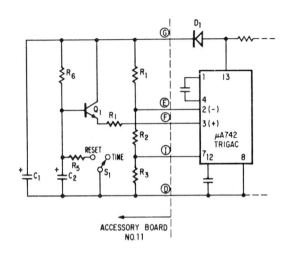

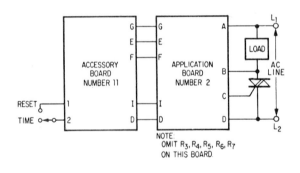

# SECTION VI
# Projects Using ICs

There are numerous applications for a 1- to 2-watt monolithic audio amplifier to drive a speaker or earphones. The sensitivity and input impedance of the GE PA-237 IC are suitable for connection to a ratio-detector (TV and FM) or a phonograph cartridge by using a resistor in series with the input to trade sensitivity for higher input impedance. For instance, with a 330K series resistor, the 2-watt sensitivity is 1-volt with the circuit shown in Fig. 6-1. Fig. 6-2 shows how cir-

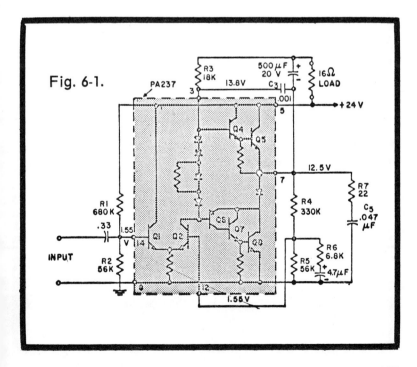

## Fig. 6-2.

| Resistor added in series with input capacitor | Input Resistance | 2-Watt Sensitivity |
|---|---|---|
| 0 | 40K | 120mv |
| 68K | 108K | 300mv |
| 120K | 160K | 450mv |
| 330K | 370K | 1.0v |
| 470K | 510K | 1.4v |
| 680K | 720K | 2.0v |

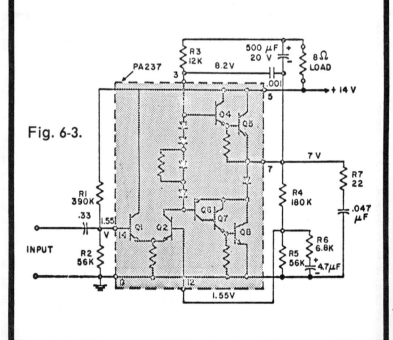

Fig. 6-3.

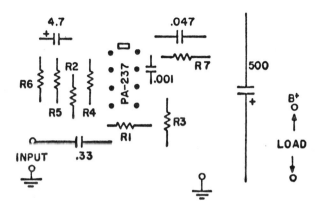

Fig. 6-4.

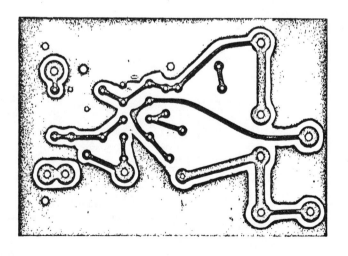

cuit sensitivity varies as the value of the resistor in series with the input capacitor is varied. It also shows the change in the input impedance of the circuit.

The PA-237 will deliver 1.5 watts into a 22-ohm load with a 22-volt power supply. With a 14-volt power supply and an 8-ohm load, the power supply is typically 1 watt with total harmonic distortion of 1% at 1000 Hz.

The circuit in Fig. 6-3 can be used with an 18-volt power supply by changing the values of R1 to 470K; R3 to 15K; and R4 to 220K. Using an 18-volt supply, the load can be either 8 or 16 ohms for about 1.5 watts power output. However, with an 8-ohm load, efficiency drops to about 40% and limits the maximum operating temperature on the package tab to 50°C for this power output.

To improve cooling, add a copper fin to the circuit board where it is soldered to the tab of the PA-237. Select an area for the fin that will maintain the package (tab) temperature within its maximum safe ratings. A pattern for a printed circuit board and parts layout are shown in Fig. 6-4.

Some applications may require that you connect one side of the load to ground potential. The circuit of Fig. 6-5 meets

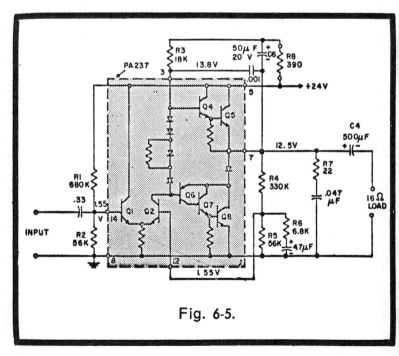

Fig. 6-5.

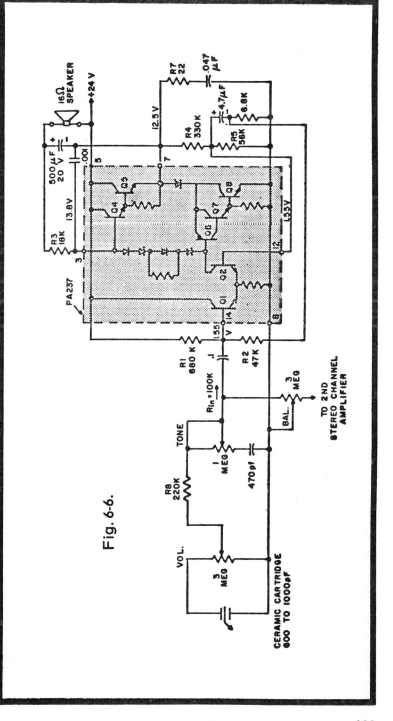

Fig. 6-6.

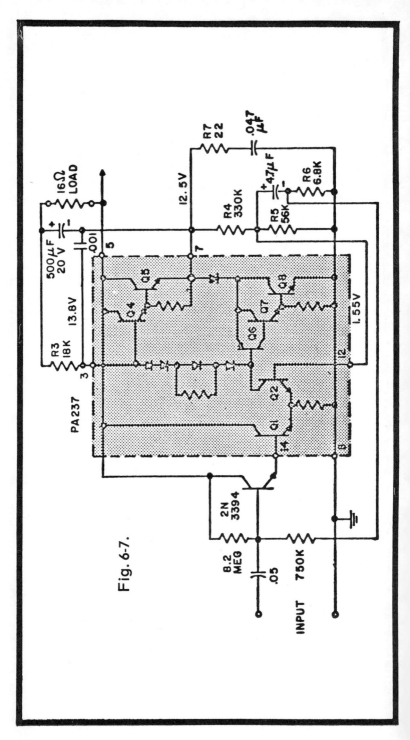

Fig. 6-7.

this requirement with the simple addition of a separate resistor and capacitor (R8 and C6) for the bootstrap (i.e., positive feedback) network. Circuit performance remains unchanged.

The input impedance of the PA-237 can be increased by bootstrapping resistor R2 as shown in Fig. 6-5, a 2-watt IC phonograph amplifier. Since the signal at both ends of resistor R2 is in phase, it increases its effective impedance. The input impedance is increased to 100K and thus decreases the loading on a ceramic cartridge and improves the bass frequency response. The phono cartridge output level will determine the optimum resistance for R8. The complete circuit appears in Fig. 6-6 and is designed for a nominal 0.7-volt output ceramic cartridge with a capacitance between 600 and 1000 pf. A minimum cartridge output as low as 0.5 volt will drive the amplifier to a full 2-watt output. With 2 watts output the total harmonic distortion at 1000 Hz is typically between 1.5 and 2%.

In normal operation, the volume control setting decreases the cartridge loading (compared to maximum output) and increases the bass frequency response. The tone control, when set for maximum treble cut attenuates a 10,000-Hz signal by 10 db or more (depending upon the volume setting) with respect to 1000 Hz. A 1000-Hz signal is changed by 1 db or less at any tone control setting.

When an input impedance of 2 or 3 megohms is required, a single transistor can be added as in Fig. 6-7. This arrangement uses the same bootstrap technique shown in Fig. 6-6, but the Darlington connection increases the total impedance.

A tape playback system is shown in Fig. 6-8. It uses a two-transistor preamp to drive the 2-watt IC power amplifier. The preamp is equalized for 1-7/8 or 3-3/4 ips tape speed with resistor R15 set at approximately 15K. The overall system voltage amplification versus frequency response is shown in Fig. 6-9.

The treble equalization control (R15) can be adjusted to compensate for variations in the program material, the tape head, or the speaker. Also, it can function as the normal treble (cut) control. The sensitivity of the system shown in Fig. 6-8 can be set by adjusting the value of R18. The circuit arrangement, as shown, will deliver 2 watts output with 0.7 mv input signal from the tape head. This order of sensitivity is adequate for an 8-track (cartridge type) stereo tape playback system. The output noise (weighted) is more than 50 db below the 2-

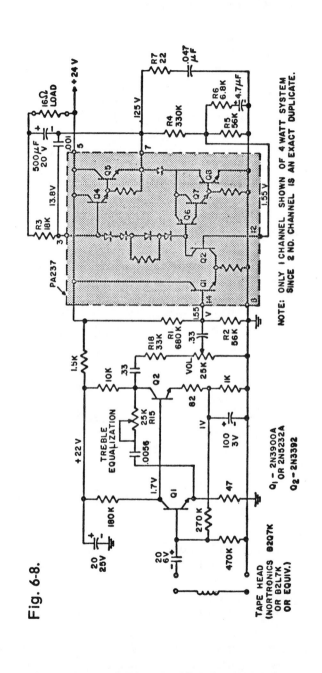

Fig. 6-8.

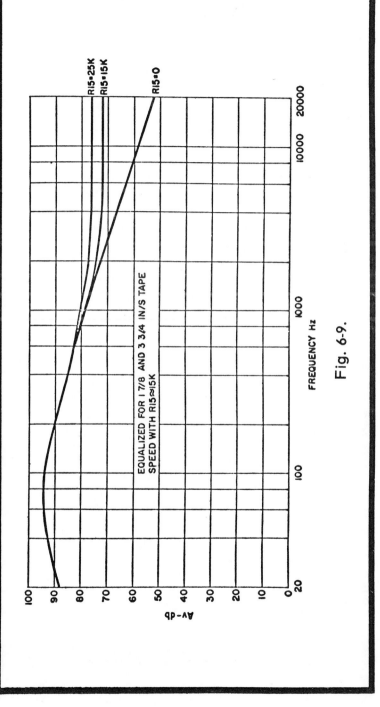

Fig. 6-9.

watt signal level. The preamplifier itself will accept input levels of 8 mv before clipping begins; this is 20 db above the input signal for 2 watts output.

## IC REGULATED POWER SUPPLIES

A laboratory-type power supply using an IC voltage regulator is shown in Fig. 6-10. It will deliver a voltage variable between 0 and 40 volts at 0 and 500 ma. The load regulation is 0.01%, and it has a remote sensing control at the plus S and minus S terminals. Although this 40-volt circuit does not flex the voltage muscles of the Motorola MC1566 integrated circuit, it does demonstrate the IC's functions.

Input frequency compensation of the auxiliary voltage supply is handled by C1 located between pins 13 and 14. For applications requiring extrenely low output noise, R2 (the voltage control) may be bypassed with C2 which should have a value between 0.1 and 2.0 mfd. Adding C2 requires that diode D1 be added between IC pins 3 and 8 to protect the IC during short circuit operation. This diode carries very little current, but should have a breakdown voltage rating that is higher than the maximum power supply output voltage. For output voltages

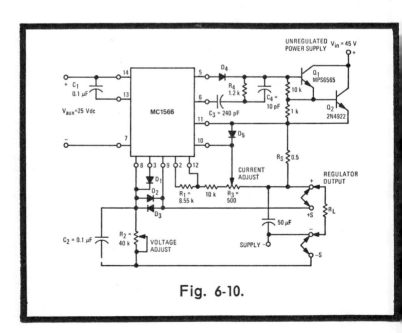

Fig. 6-10.

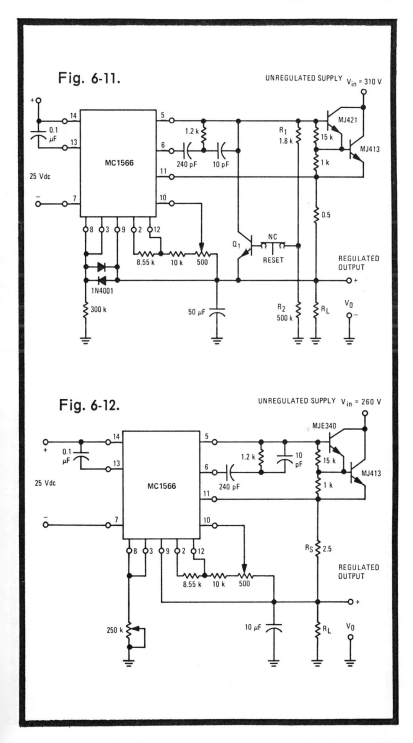

Fig. 6-11.

Fig. 6-12.

higher than 20 volts, diodes D2 and D3 should be added to protect the voltage differential amplifier.

A higher voltage power supply is shown in Fig. 6-11. This circuit is a 300-volt, 500 ma regulated supply. The circuit includes provision for automatic shutdown to protect the series-pass transistors during overload conditions.

Still another regulated supply is shown in Fig. 6-12. This one delivers 0 to 250 volts at 100 ma. Notice that the voltage-setting resistor is variable to permit the selection of an output voltage from nearly zero to full output rating of the supply.

The last power supply built around this IC is shown in Fig. 6-13. In this arrangement the power output transistors are run close to their limits. The circuit will deliver 100 volts at 200 ma. If the output is shorted accidentally, however, the series-pass transistors will be destroyed. The IC will survive as it is protected against damage.

## 1-WATT MONOLITHIC IC AMPLIFIER

Here is a complete IC amplifier that can deliver 1 watt of audio power to an 8-ohm load when powered by a 9-volt supply.

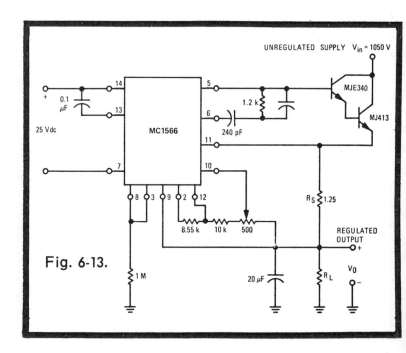

Fig. 6-13.

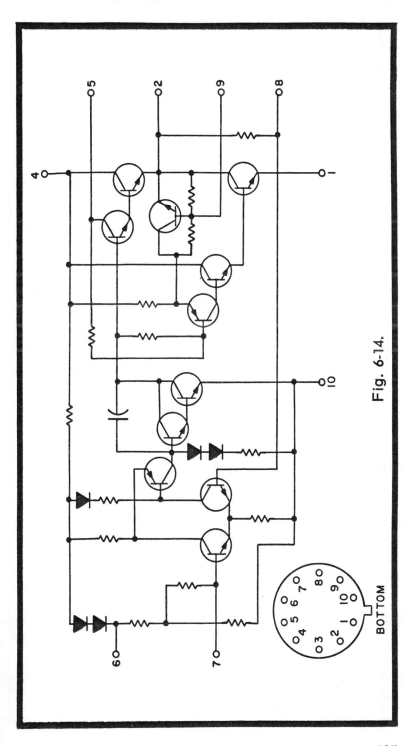

Fig. 6-14.

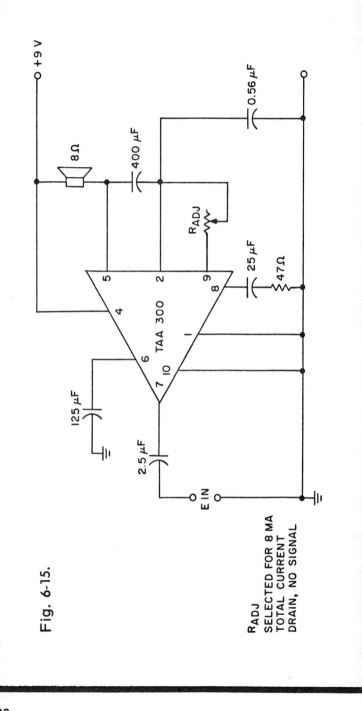

Fig. 6-15.

RADJ SELECTED FOR 8 MA TOTAL CURRENT DRAIN, NO SIGNAL

High gain and low current consumption make the Amperex TAA300 suitable for use as a phonograph amplifier, radio or TV audio amplifier, walkie-talkie audio and speech amplifiers, tape cassette systems, and mini hi-fi systems.

The internal construction of the device is shown in Fig. 6-14. In Fig. 6-15 we show a typical audio amplifier using the TAA300 to feed an 8-ohm load. Fig. 6-16 is set up for a 4-ohm load with the speaker connected to the positive line. Fig. 6-17 is for an 8-ohm load with the speaker in the ground circuit.

You will notice that in all instances a minimum of external circuit elements is required and, therefore, each of these circuits can be built into a very small space.

## USING AN IC DIODE ARRAY

Here is a new IC that contains six diodes. Four of them are arranged in a diode-quad arrangement and the other two are individual units. The internal arrangement of this RCA CA-3019 IC is shown in Fig. 6-18.

Each diode in the IC is formed from what could have been a transistor by connecting the collector and base elements together to form the diode anodes, while the emitters are used as the diode cathode. (Incidentally, this is only one of five ways to convert a transistor into a diode.)

Fig. 6-19 shows a circuit for a typical synthesizer mixer. Shorting IC Terminals 5 and 8 provides two independent sets of back-to-back diodes which are useful for limiting and clipping. Fig. 6-20 is a limiter circuit using the diodes in this way.

Fig. 6-21 shows how the CA3019 can be used as a balanced-modulator that minimizes the carrier frequency from the output with a symmetrical bridge network. A carrier of the opposite polarity causes all the diodes to conduct and thus effectively short-circuits the signal source. A carrier of the opposite polarity cuts off all the diodes and allows signal current to flow to the load. If the four diodes are identical, the bridge is perfectly balanced and no carrier current flows in the output load.

In high-speed gates, the gating signal often appears at the output and causes the output signal to ride a "pedestal." A

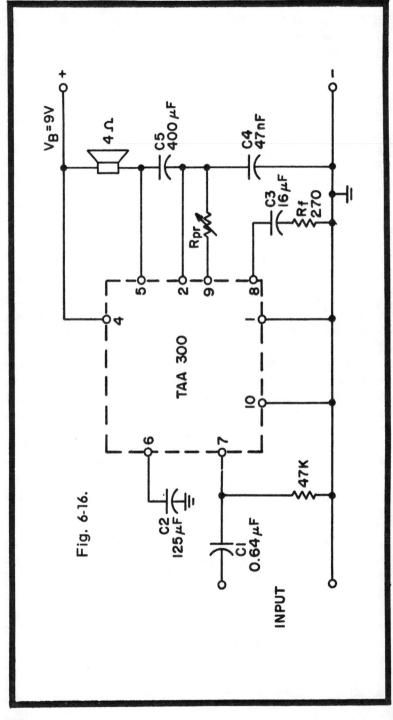

Fig. 6-16.

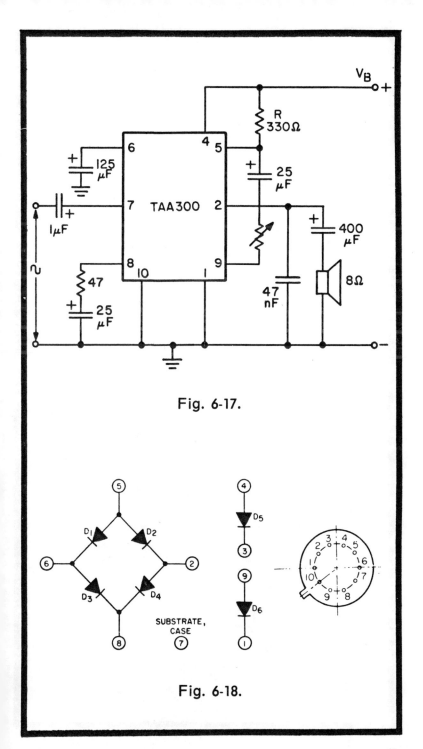

Fig. 6-17.

Fig. 6-18.

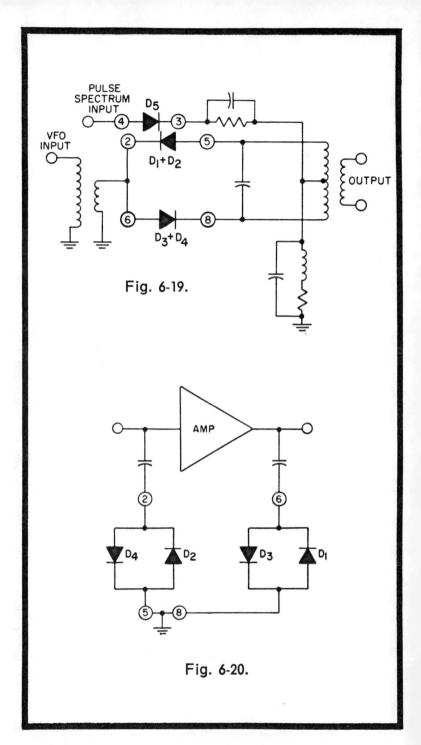

Fig. 6-19.

Fig. 6-20.

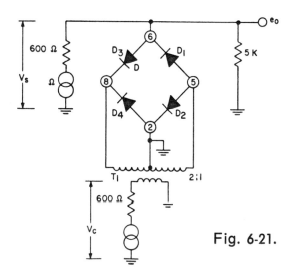

Fig. 6-21.

$T_1$ — Technitrol No. 8511660 or equiv.

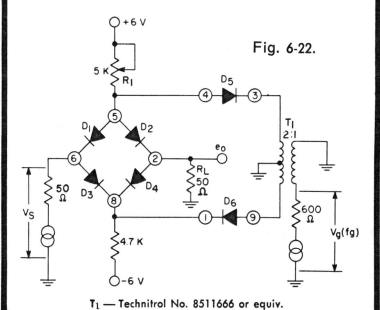

Fig. 6-22.

$T_1$ — Technitrol No. 8511666 or equiv.

diode-quad bridge circuit can be used to balance out the undesired gating signal at the output and reduce the pedestal to the extent that the bridge is balanced.

A diode-quad gate acts as a variable impedance between a source and a load and can be connected either in series or in shunt with the load. The circuit configuration used depends on the input and output impedances of the circuits to be gated. A series gate is used if the source and load impedances are high compared to the diode forward resistance.

Series Gate

Fig. 6-22 shows how to use the CA3019 as a series gate in which the diode bridge, in series with the load resistance, balances out the gating signal to provide a pedestal-free output. With a proper gating voltage (1 to 3 volts RMS, 1 to 500 kHz), diodes D5 and D6 conduct during one half of each gating cycle and do not conduct during the other half of the cycle. When diodes D5 and D6 are conducting, the diode bridge (D1 through D4) is not conducting and the high diode back resistance prevents the input signal Vs from appearing across load resistance RL. When diodes D5 and D6 are not conducting, the diode bridge conducts and the low diode forward resistance allows the input signal to appear across the load resistance. Resistor R1 may be adjusted to minimize the gating voltage at the output.

Shunt Gate

Fig. 6-23 shows how the RCA CA3019 can be used as a shunt gate. In this arrangement, the diode bridge in shunt with the load resistance balances out the gating signal to provide a pedestal-free output. When gating voltage Vg is high enough, the diode bridge (D1 through D4) conducts during one half of each gating cycle and does not conduct during the other half of the cycle.

When the diode bridge is conducting, its low diode forward resistance shunts load resistance RL and prevents input signal Vs from appearing at the output. When the diode bridge

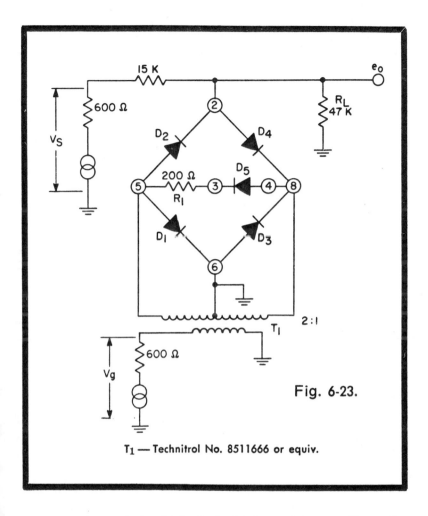

Fig. 6-23.

$T_1$ — Technitrol No. 8511666 or equiv.

is not conducting, its high diode back resistance allows the input signal to appear at the output. Diode D5 and R1 keep the transformer load nearly constant during both halves of the gating cycle.

Series-Shunt Gate

A series-shunt gate that uses all six diodes of the CA3019 is shown in Fig. 6-24. This circuit arrangement combines the good on-to-off impedance ratio of the shunt gate with the low-output pedestal of the series gate.

On the gating half cycle during which the voltage at A is positive with respect to the voltage at B, there is no output because the shunt diodes are forward biased and the series diodes are reverse biased. Any signal passing through the input diodes (D4 and D2) runs into a low shunt impedance to ground (D5 and D6) and a high impedance in series with the signal path to the load (D3 and D1). This circuit assures a good on-to-off impedance ratio. When the voltages at A and B reverse, the conduction states of the shunt and series diodes reverse, and the signal passes through the gate to load resistor RL. Any pedestal at the output is a function of the resistor, transformer, and diode balance.

The gate will continue to operate successfully with R1 and R2 shorted if the transformer center tap is removed from ground. In either case, no DC supply is required to bias the gate diodes.

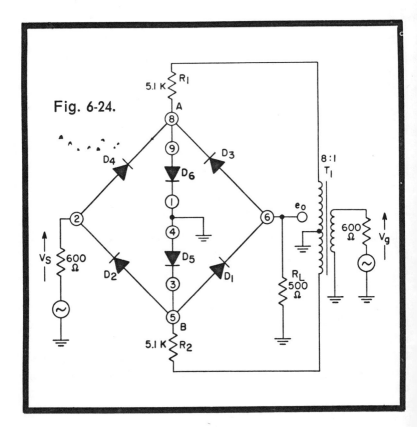

Fig. 6-24.

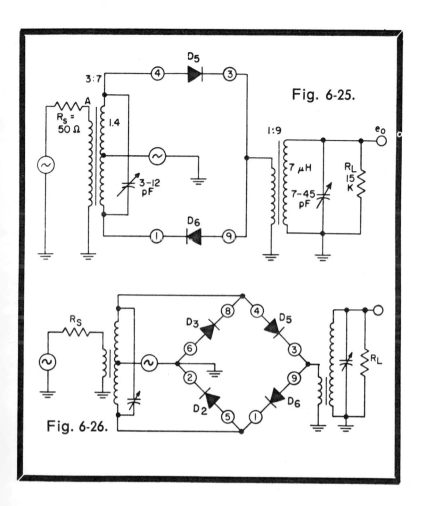

Fig. 6-25.

Fig. 6-26.

## Balanced Mixer

Fig. 6-25 shows how to use the CA3019 as a conventional balanced mixer. The load resistor across the output tuned circuit is selected to provide maximum power output.

The CA3019 mixer shown in Fig. 6-26 is essentially a balanced mixer with two additional diodes (D3 and D4) added to form a half-wave carrier switch. The additional diodes permit both legs of the circuit D1-D2 and D3-D4 to operate throughout the AC cycle. As compared with the conventional balanced mixer shown in Fig. 6-25, this circuit effectively doubles the desired output voltage and reduces the output voltage at the oscillator frequency by half. However, the capacitances associated with the integrated diodes prevent this circuit arrangement from realizing the improvement in conversion gains at frequencies above 20 MHz.

Ring Modulator

In Fig. 6-27 we show the CA3019 used as a ring modulator. If a perfectly balanced arrangement were used, carrier current of equal magnitude and opposite direction would flow in each half of the center-tapped transformer T2. Thus the effect of the carrier current in T2 would be cancelled and the carrier frequency would not appear in the output. However, the ring modulator is not perfectly balanced because diodes D1 and D2 and D3 and D4 are actually two diodes in parallel, while diodes D5 and D6 are individual diodes. Nevertheless, this circuit attenuates the carrier in the output as well as an arrangement that uses both individual diodes in two CA3019 circuits.

As the carrier passes through half of its cycle, diodes D1, D2 and D6 conduct and diodes D3, D4 and D5 do not conduct. When the carrier passes through the other half of its cycle, the

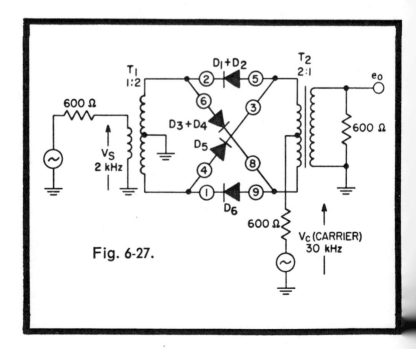

Fig. 6-27.

TABLE 6-1    PERFORMANCE CHARACTERISTICS OF RING MODULATOR

FOR A GIVEN Vs + Vg' eo IN MILLIVOLTS

| OUTPUT FREQ. (kHz) | | | | | | |
|---|---|---|---|---|---|---|
| | Vs (mv) | 300 | 350 | 450 | 500 | |
| | Vc (mv) | 600 | 500 | 350 | 300 | |
| 28 OR 32 | UPPER OR LOWER SIDEBANDS | 86 | 97 | 83 | 91 | |
| 2 | SIG. FREQ. | 0.042 | 0.02 | 0.015 | 0.020 | |
| 30 | CARRIER FREQ. | 1.3(-37db)* | 0.88(-41db)* | 0.67(-42db)* | 0.62(-43db)* | |
| 26 or 34 | HIGHER ORDER SIDEBANDS | 0.018 | 0.016 | 0.036 | 0.043 | |
| 24 or 36 | | 0.021 | 0.054 | 0.047 | 5.0 | |

*db BELOW THE DESIRED UPPER AND LOWER SIDEBANDS

previously nonconducting diodes (D3, D4 and D5) conduct, and vice versa. As a result, the output amplitude is alternately switched from plus to minus at the carrier frequency. The signal frequency component of the output waveform is thus symmetrical about the zero axis and is not present in the output. Therefore, the ring modulator suppresses both the carrier frequency and the signal frequency so that the output theoretically contains only the upper and lower sidebands. For single-sideband transmission, one of these sidebands can be eliminated by selective filtering. The performance of the CA3019 as a ring modulator is shown in Table 6-1.

DIGITAL ORGAN TONE GENERATOR

Here is a digital tone generator that divides a single frequency, generated by a crystal-controlled oscillator, to produce a full chromatic scale for organ music. Such a system has several advantages over conventional electronic organs. Of prime importance is that this kind of organ can never go out of tune. And of even more interest to the experimenter is that such a tone generator is easy to build and no tuning is required.

The musical scale, which is usually seven to eight octaves long, is a group of related frequencies. For any given note with a frequency of f, there are harmonically related notes, called octaves, that have frequencies of 2f, 4f, 8f, 1/2f, 1/4f, 1/8f, 1/16f, and so on. There are 12 notes in each octave. Thus, the frequency ratio of any two adjacent halftones is the 12th root of 2, which is an irrational number.

The ratio of the whole numbers 196 and 185 equals the value of the 12th root of 2 to the first six digits. With the ratio of these two numbers, several high-quality electromechanical tone generators have been built. The best example of such a device is the Hammond tone-wheel organ, in which one shaft drives another from an 185-tooth gear into a 196-tooth gear. This sequence, carried through 12 such shafts equipped with tone wheels of continually doubling teeth (2, 4, 8, 16, 32, etc.) provides all of the fundamental musical frequencies accurately (better than 10 ppm). Such a generator could be implemented electronically by taking some stable high frequency, dividing it by 196, multiplying it by 185, and continuing this

### DIVISORS FOR DIGITAL TONE GENERATOR
(Oscillator Frequency: 774.4 kHz)

| Note | Divisor | True-Scale Frequency, Hz | Digital Frequency, Hz | Approx Error, Cents* |
|------|---------|--------------------------|-----------------------|----------------------|
| C    | 185     | 4186.01                  | 4186                  | 0                    |
| B    | 196     | 3951.07                  | 3951                  | 0                    |
| A#   | 208     | 3729.31                  | 3723                  | -3                   |
| A    | 220     | 3520.00                  | 3520                  | 0                    |
| G#   | 233     | 3322.44                  | 3324                  | +1                   |
| G    | 247     | 3135.96                  | 3135                  | 0                    |
| F#   | 261     | 2959.96                  | 2967                  | +4                   |
| F    | 277     | 2793.83                  | 2796                  | +2                   |
| E    | 294     | 2637.02                  | 2634                  | -3                   |
| D#   | 312     | 2489.02                  | 2482                  | -4                   |
| D    | 330     | 2349.32                  | 2347                  | -1                   |
| C#   | 350     | 2217.46                  | 2213                  | -3                   |

*One cent is 1% of a halftone frequency.

TABLE 6-2

process through a total of 12 sequences. The only problem with this method is that while it is easy enough to divide frequencies by 196, it is very difficult, if indeed not impossible, to accurately multiply frequencies by 185. Therefore, the system proposed here would use only dividing.

As stated earlier, 185 and 196 are very good dividing numbers, called divisors. (The term "divisor" is used here to distinguish from "divider" which ordinarily means dividing by 2 in organ language.) Any frequency divided by both 185 and 196 will produce two tones precisely a halftone apart, to an accuracy of 0.0001%. The next number in the series is 208 which, when it is divided by 196, also gives the value of the 12th root of 2, but the resulting number is less accurate by two orders of magnitude. The next number in the series is 220. Continuing this process, 12 such numbers can be found, as shown in Table 6-2, which can be used to produce a fairly accurate musical scale.

One way to get a more accurate scale, at the expense of expanded digital circuitry, is to use larger divisors. Because of clock speed limitations in some ICs, an expansion ratio of 5 was chosen for use here. This ratio provides a 3.872-MHz clock rate, which is well within the specified clock rate of the MRTL ICs we use, and provides a scale whose maximum frequency deviation is plus or minus 0.6 cent. One cent on this scale is 1% of a halftone frequency, which represents very accurate tuning. These new divisors and the calculated errors are in Table 6-3.

The overall system, in which the single 3.872-MHz fre-

## DIVISORS FOR DIGITAL TONE GENERATOR

(Oscillator Frequency: 3.872 MHz)

| Note | Divisor | True-Scale Frequency, Hz | Digital Frequency, Hz | Approx Error, Cents* |
|------|---------|--------------------------|------------------------|----------------------|
| C8   | 925     | 4186.01                  | 4186.0                 | 0                    |
| B7   | 980     | 3951.07                  | 3951.1                 | 0                    |
| A#7  | 1038    | 3729.31                  | 3730.4                 | +0.5                 |
| A7   | 1100    | 3520.00                  | 3520.0                 | 0                    |
| G#7  | 1165    | 3322.44                  | 3323.5                 | +0.55                |
| G7   | 1235    | 3135.96                  | 3135.3                 | -0.3                 |
| F#7  | 1308    | 2959.96                  | 2960.3                 | +0.1                 |
| F7   | 1386    | 2793.83                  | 2793.6                 | -0.14                |
| E7   | 1468    | 2637.02                  | 2637.0                 | 0                    |
| D#7  | 1555    | 2489.02                  | 2490.0                 | +0.5                 |
| D7   | 1648    | 2349.32                  | 2349.7                 | +0.2                 |
| C#7  | 1746    | 2217.46                  | 2217.8                 | +0.2                 |
| C7** | 1849    | 2093.0                   | 2094.0                 | +0.5                 |

*One cent is 1% of a halftone frequency.

**This divisor is not used in the tone generator. It was applied for the calculation only to see how well the octave will close. Note that the octave closes to within 1 Hz, or 0.5 cent.

TABLE 6-3

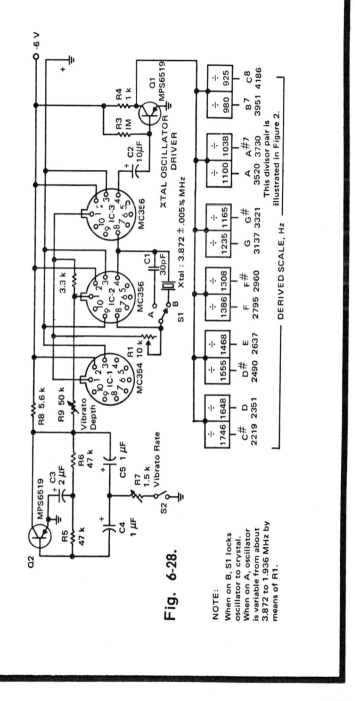

Fig. 6-28.

NOTE:
When on B, S1 locks oscillator to crystal.
When on A, oscillator is variable from about 3.872 to 1.936 MHz by means of R1.

quency is divided into 12 related frequencies, is shown in logic block diagrams in Figs. 6-28 and 6-30. The 12 divisor systems are driven from the oscillator simultaneously and provide 12 halftones with frequencies ranging from 4186 to 2217.5 Hz, with no more than plus or minus 2 Hz worse-case error. These 12 outputs are routed into 12 divide-by-two dividers to provide a 7-octave, complete chromatic scale.

Let us take, for example, the numbers 1038 and 1100, which appear as the third and fourth entries in Table 6-3. The objective is to design this part of the system so that it will divide the oscillator frequency by 1038 and 1100. Dividing by each of these numbers requires 11 binary counters, implemented with five 1/2 MC790P dual MRTL J-K flip-flops. Other parts required are an MC715P 3-input gate, two MC-786 dual 4-input expanders, and an MC788P dual buffer driver stage. Since the gate, expanders, and the buffer are dual units, one set of these can serve two divisor systems. Thus both the 1038 and 1100 divisor systems are considered here.

The only difference between these is the connection of the counter outputs to the gate inputs. The cascaded binary counter accumulates counts in straight binary fashion. When the critical count is reached, as determined by the logic connections to the gate inputs, the gate produces a signal, which, through the buffer, resets all counter stages to the zero state, and the count starts all over again.

The dual divisor stage is set up as indicated in Fig. 6-29. The IC elements are shown as logic blocks. (The numbers shown are the pin numbers for the 14-pin dual in-line package.) The logic connections for any of the divisors can be determined as follows:

First the divisor is converted to binary. The most convenient way of doing this is to enter the divisor above the largest binary place number that will subtract from the divisor. The remainder is entered above the largest binary place number that will subtract from it. This process is repeated until complete, as shown in Table 6-4.

Whenever a subtraction can be made, a binary 1 is entered; at other places a binary 0. Then the binary numbers are converted to their complements. The flip-flops are designated in alphabetical order from the head of the chain, and where a binary 1 appears in the complement, a gate input is connected

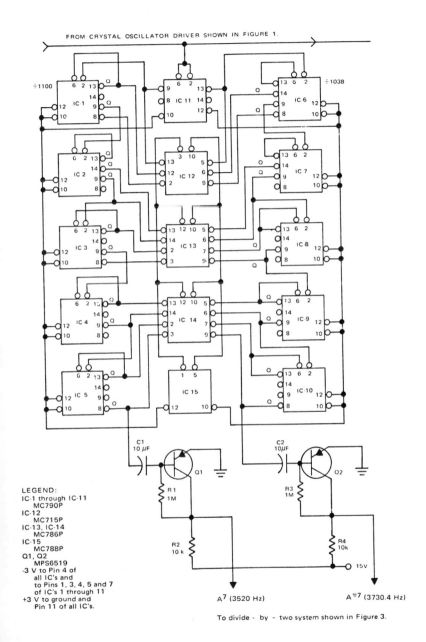

Fig. 6-29.

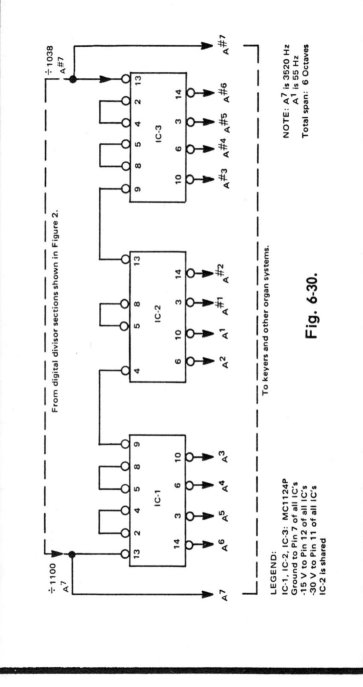

**Fig. 6-30.**

to the flip-flop Q output. Where a binary 0 appears, a gate input is connected to the flip-flop $\overline{0}$ output.

All of the divisor pairs are made up in this fashion. The first two numbers, 925 and 980, require only 10 flip-flops; all the rest, 11. All pairs require one dual 3-input gate, two dual gate expanders, and one dual buffer. Some of the gate or expander inputs are not used. These should be grounded to prevent instabilities. This digital system requires 90 MRTL ICs. On a single-circuit basis, the divisor logic requires 178 circuit functions.

The 12 digital outputs drive a system of divide-by-two binary elements that are equivalent to the dividers in a conventional organ divider. The MC1124P MOS dividers were designed by Motorola especially for organ use. Part of the divide-by-two system is shown in Fig. 6-30.

For driving the MOS elements from the MRTL outputs, an interface element (a transistor drive stage) is required as shown in Fig. 6-29 (Q1 and Q2). These 12 drivers and the oscillator driver are the only nonintegrated elements in the entire tone generator system.

The MOS divide-by-two section requires 68 flip-flops. Since there are four flip-flops in a package, this necessitates 17 Motorola MC1124P divider packages. The overall system provides 80 chromatic intervals, or almost seven full octaves.

In the experimental organ featuring the digital tone generator system, a dual purpose oscillator is used. One of the oscillator modes is crystal locked at 3.872-MHz. The other is variable for approximately one octave, from 3.872 MHz down. The crystal oscillator locks the whole system to the Inter-

---

Divisor = 1100

| Place Number | 1 | 2 | 4 | 4<br>8 | 12<br>16 | 32 | 76<br>64 | 128 | 256 | 512 | 1100<br>1024 |
|---|---|---|---|---|---|---|---|---|---|---|---|
| | – | – | 0 | 4 | – | – | 12 | — | — | — | 76 |
| Binary Number | 0 | 0 | 1 | 1 | 0 | 0 | 1 | 0 | 0 | 0 | 1 |
| Complement | 1 | 1 | 0 | 0 | 1 | 1 | 0 | 1 | 1 | 1 | 0 |
| Connections | A | B | C | D | E | F | G | H | I | J | K |

TABLE 6-4

national Pitch Musical Scale. The variable oscillator enables the system to be pitched to any arbitrary frequency within an octave. Operating on the fixed oscillator, the system will never drift, except for very small changes caused by temperature changes in the crystal. Operating on the variable oscillator, some of the features available include tuning ability (to tune the organ to other instruments which may not be precisely up to standard pitch), a very uniform gliding tone and vibrato, and key-shift capability.

The oscillator, built around a Motorola MC356 MECL IC gate is shown in Fig. 6-28. One advantage of this IC is that it provides an amplifier in which the output and input are in phase, resulting in some very simple oscillator configurations now requiring a $180^\circ$ phase shift feedback.

IC1 is an MC354 bias driver which furnishes Vbb bias for IC2 and IC3. IC3 acts as a buffer stage, and Q1 is an interface amplifier to adjust the MECL logic level to that required by the MRTL divisors. Q2 and the associated RC network provide a twin-T oscillator with a frequency of about 6 Hz. This signal is injected to pin 1 of IC2 through R9, providing an excellent vibrato (frequency modulation) effect when the oscillator is on crystal mode. However, if the vibrato is not effective when the oscillator is on crystal mode, it can be turned off with switch S2.

## USING THE RCA NUMITRON

The RCA DR2000, DR2010, DR2020, and DR2030 Numitrons are incandescent digital display devices that provide sharp, high brightness displays ideally suited for most types of digital readout systems. These devices, which may be operated in either a DC or multiplex mode, offer the designer of digital-display systems several advantages which include:

1. High contrast clutter-free displays in virtually any desired color with controllable brightness.
2. Low-voltage operation (4.5 volts nominal).
3. Wide viewing angle.
4. Minimum center-to-center mounting distance between adjacent devices of only 0.80 inch.

**DR2000**
0 through 9

**DR2010**
0 through 9
with decimal point

**DR2020**
Plus-Minus sign
and numeral 1

**DR2030**
Plus-Minus sign

Fig. 6-31.

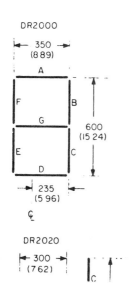
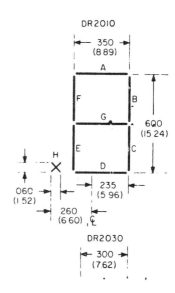

Fig. 6-32.

TABLE 6-5

## PIN CONNECTIONS AND COIL SEGMENTS USED TO FORM DIGITAL-CHARACTER DISPLAY ON RCA NUMITRONS

| Display | Device Pin Designation | | | | Corresponding Coil Segments | | | |
|---|---|---|---|---|---|---|---|---|
| | | | | | Pin No.2 Common For All Types | | | |
| | DR2000 | DR2010 | DR2020 | DR2030 | DR2000 | DR2010 | DR2020 | DR2030 |
| ⌐⌐ | 3,4,5,7,8,9 | 3,4,5,7,8,9 | | | E,D,C,A,B,F | E,D,C,A,B,F | | |
| − | 5,8 | 5,8 | 6,8 | | C,B | C,B | D,C | |
| ⌐⌐ | 3,4,6,7,8 | 3,4,6,7,8 | | | E,D,G,A,B | E,D,G,A,B | | |
| ⌐⌐ | 4,5,6,7,8 | 4,5,6,7,8 | | | D,C,G,A,B | D,C,G,A,B | | |
| ⌐⌐ | 5,6,8,9 | 5,6,8,9 | | | C,G,B,F | C,G,B,F | | |

| | | | | | | |
|---|---|---|---|---|---|---|
| ⊔ | 4,5,6,7,9 | 4,5,6,7,9 | | D,C,G,A,F | D,C,G,A,F | |
| ⊔ | 3,4,5,6,7,9 | 3,4,5,6,7,9 | | E,D,C,G,A,F | E,D,C,G,A,F | |
| ⊓ | 5,7,8 | 5,7,8 | | C,A,B | C,A,B | |
| ⊔ | 3,4,5,6,7,8,9 | 3,4,5,6,7,8,9 | | E,D,C,G,A,B,F | E,D,C,G,A,B,F | |
| ⊔ | 4,5,6,7,8,9 | 4,5,6,7,8,9 | | D,C,G,A,B,F | D,C,G,A,B,F | |
| (+) | | | 7,9 | 6,8 | | | B,A | B,A |
| (−) | | | 7 | 6 | | | E | B |
| (×) | | 1 | | | H | | |

5. Uses standard low-cost sockets.
6. Freedom from induced or radiated interference.
7. Full compatibility with low-cost IC decoder/drivers.
8. High reliability (life expectancy exceeds 100,000 hours).

Fig. 6-31 is an external view of the RCA Numitrons. Each device contains a number of incandescent single-helical coil segments in an evacuated glass envelope. The desired display is obtained by applying DC voltages to the appropriate coil-segment pin connections. During operation, the coil segments glow in a clear brilliant fine line. If a broader line is desired, etched or other specially treated glass can be placed in front of the device.

Fig. 6-32 shows schematic representations and dimensions of the coil-segment arrangement for each type of Numitron. The DR2000 can readout as any numeral from 0 through 9; the DR2010 can display any of the numerals from 0 through 9 and a decimal point; the DR2020 can be used to form the numeral 1 or the numeral 1 preceded by either a plus or minus sign; and the DR2030 can provide either a plus or a minus sign. Table 6-5 shows the various numerical and mathematical characters and indicates the corresponding coil segment and external pin connections to which voltages must be applied to form each character. Individual coil segments of the DR2000 and DR2010 Numitrons may also be addressed to provide other symbols, and certain alphabetical displays. Table 6-6 shows ten nonconflicting letters that can be displayed on these devices and lists the corresponding coil segment and external pin connection to which voltages must be applied to form each of them.

Characters displayed on RCA Numitrons are shown in a height-to-width aspect ratio of about 2 to 1. If larger characters are required a Fresnel lens can be used to provide the needed magnification.

During normal Numitron operation, voltages applied to individual coil segments to form the desired character are controlled by a decoder/driver circuit. One end of each coil segment in a Numitron is internally connected to a common lead. The external connection (pin 2) for this lead is usually the positive terminal for the segment voltage. The ground return (or negative connection) for the segment-voltage supply is completed through the decoder/driver.

Two RCA 16-lead IC decoder/drivers, types CD2500E and CD2501E, supplied in dual-in-line plastic packages, are available for use with the Numitrons. Each type of decoder/driver accepts four inputs in BCD 8-4-2-1 code and provides 7-segment decoded outputs that represent a numeral from 0 to 9. Table 6-7 lists the BCD inputs and indicates the coil segments of the DR2000 that are illuminated to form each of the numerals from 0 to 9. The decoder/drivers differ in that the CD2500E contains a decimal driver and the CD2501E has a ripple-blanking feature that automatically eliminates all insignificant zeros in a Numitron multi-digit display. Fig. 6-33 shows the terminal-connection diagrams for both IC decoder/drivers.

The glass envelope used for RCA Numitrons has a standard miniature 9-pin base. Therefore, these devices can be mounted in sockets normally used for miniature tubes. Numitrons may also be mounted on printed-circuit boards with direct soldering or with a socket as in Fig. 6-34. Fig. 6-34 shows a printed circuit layout for mounting a Numitron directly on a circuit board. The arrangement also shows the mounting arrangement for the CD2501E driver/decoder and the required interconnection wiring. This view is actual size from the foil side of the board. Fig. 6-34B shows a layout for mounting a Numitron in a socket and a CD2501E decoder/driver on a circuit board.

Fig. 6-35 is a simple test circuit that may be used to determine the current and luminance of individual coil segments as a function of segment voltage. Figs. 6-36 and 6-37 show the variations in these quantities with changes in the segment voltage.

Fig. 6-38 shows the basic interconnections of a Numitron with an IC decoder/driver for operation from a fixed DC supply. During circuit operation, 0.3 to 0.5 volt is dropped across the decoder/driver, so a 5-volt DC supply is required to obtain proper Numitron brightness.

Fig. 6-39 is a schematic of a simple series voltage-regulator that may be used to control Numitron brightness. The circuit operates from a 7-volt DC input and provides a variable DC output from about 2.5 to 5 volts.

To improve reliability it is desirable to use one decoder/driver for each Numitron readout. However, when a display requires more than six readouts, using one decoder/driver to multiplex the readouts can reduce cost considerably.

## TABLE 6-6

### PIN CONNECTIONS AND COIL SEGMENTS USED TO FORM ALPHABETICAL DISPLAYS ON

| Display | Device Pin Designation | Corresponding Coil Segments |
|---|---|---|
| | Pin No. 2 Common | |
| A | 3,5,6,7,8,9 | E,C,G,A,B,F |
| C | 3,4,7,9 | E,D,A,F |
| E | 3,4,6,7,9 | E,D,G,A,F |
| F | 3,6,7,9 | E,G,A,F |
| H | 3,5,6,8,9 | E,C,G,B,F |
| J | 3,4,5,8 | E,D,C,B |
| L | 3,4,9 | E,D,F |
| P | 3,6,7,8,9 | E,G,A,B,F |
| U | 3,4,5,8,9 | E,D,C,B,F |
| y | 4,5,6,8,9 | D,C,G,B,F |

TABLE 6-7

## DECODER INPUTS REQUIRED TO FORM NUMERICAL CHARACTERS ON THE DR2000 NUMITRON

| Display | BCD INPUT (8-4-2-1 CODE) TO DECODER/DRIVER | | | |
|---|---|---|---|---|
| | D (TERM. 6) | C (TERM. 2) | B (TERM. 1) | A (TERM. 7) |
| 0 | 0 | 0 | 0 | 0 |
| 1 | 0 | 0 | 0 | 1 |
| 2 | 0 | 0 | 1 | 0 |
| 3 | 0 | 0 | 1 | 1 |
| 4 | 0 | 1 | 0 | 0 |
| 5 | 0 | 1 | 0 | 1 |
| 6 | 0 | 1 | 1 | 0 |
| 7 | 0 | 1 | 1 | 1 |
| 8 | 1 | 0 | 0 | 0 |
| 9 | 1 | 0 | 0 | 1 |

**Notes:** 1. D, C, B, and A represent the BCD code 8-4-2-1, respectively, which are the four input to the decoder/driver required to illuminate the corresponding digit on the Numitron.

2. 0 = low-level input; 1 = high-level input.

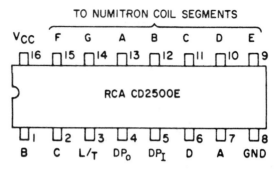

$DP_O$ = Decimal Point Output
$DP_I$ = Decimal Point Input
$DP_I$ must be supplied from an external source
The CD2500E performs the inverter-driver function necessary to energize the decimal point filament in the display device.

(a)

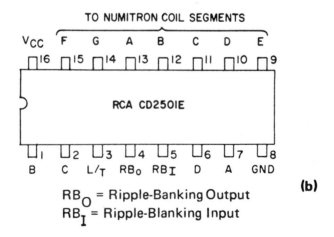

$RB_O$ = Ripple-Banking Output
$RB_I$ = Ripple-Blanking Input

(b)

Fig. 6-33.

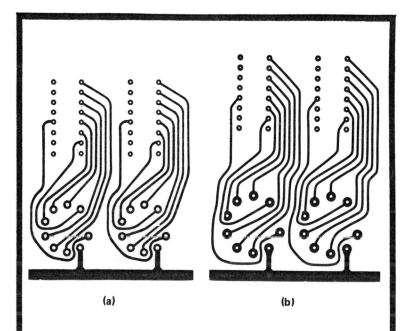

Fig. 6-34.

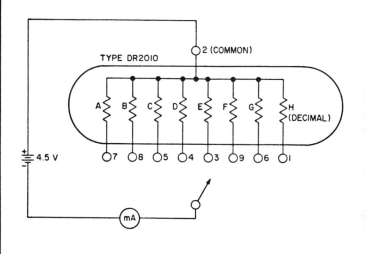

Fig. 6-35.

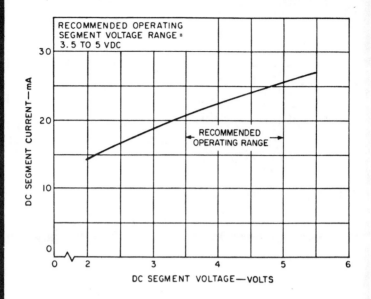

Fig. 6-36.

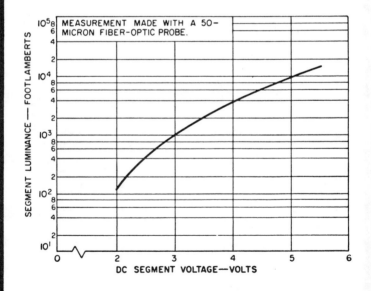

Fig. 6-37.

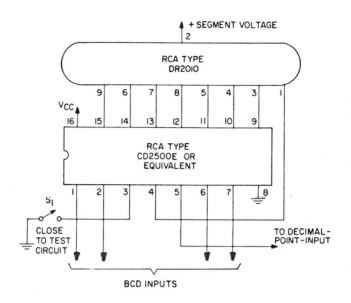

Fig. 6-38.

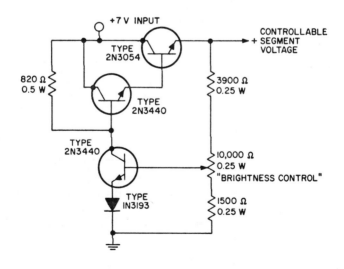

Fig. 6-39.

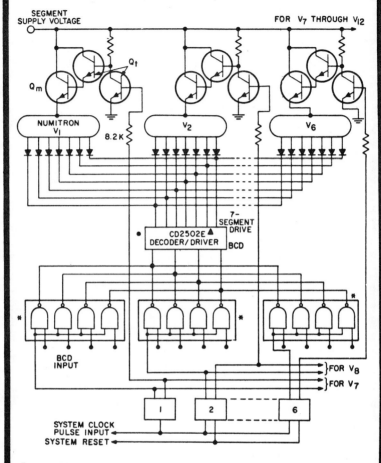

\* CD2302 or equivalent

▲ Selection of the CD2502E rated for operation at 12 volts.

● The CD2502E can drive up to six Numitrons. Eight Numitrons may be multiplexed provided that a suitable decoder/driver (rated for operation at 15 volts and 60 milliamperes) is used. Required decoder/driver ratings are determined from Fig. 14.

Fig. 6-40.

One method of multiplex operation is shown in Fig. 6-40. Here, one RCA CD2502E decoder/driver is used to drive up to a maximum of six DR2000 Numitrons with a duty factor of 16.2%. When DR2010 Numitrons are used, a separate drive circuit for the decimal points must be included. The multiplex circuit uses a ring counter, which in sequence controls the BCD data in the NAND gates and the application of the segment-voltage pulses, which pass through drive transistor $Q_m$ to the common terminal of each Numitron. In this way, the ring counter determines which device in the series will provide the proper numerical display at a given time. The illumination of the proper coil segment to form the desired numeral on each Numitron is still controlled by the decoder/driver in response to the BCD coded inputs. Isolation diodes are needed in series with each coil segment of the Numitron to prevent simultaneous lighting of coil segments in adjacent devices.

By far, the most common digital measuring instruments are the digital counter timer and the digital voltmeter (DVM). A digital counter using Numitrons is shown in Fig. 6-41. A ramp type digital voltmeter is in Fig. 6-42. In the ramp method, the DC signal is compared against an accurate ramp. The width of the output signal is proportional to the DC voltage being measured. The comparator output operates a gate that allows a number of oscillator cycles through the gate in direct proportion to the input voltage. With 1 volt at the input, exactly 1000 cycles pass through the gate and are totaled on the counters. Similarly, an input of 0.5 volt results in exactly 500 counts.

## AN IC STEREO PREAMP

One channel of a stereo preamp is shown in Fig. 6-43. The Motorola MC1303P IC, shown schematically in Fig. 6-44 is a dual preamplifier in a single case and is intended for use in stereo amplification. Each channel of the amplifier has a differential input amplifier, followed by a second differential stage with single-ended output and two emitter-follower stages.

The input differential amplifier is fed from a constant-current source in the emitters which, in turn, is biased from a voltage divider in the emitter circuit of the second stage differential amplifier. This arrangement is used because it provides common-mode negative feedback to increase rejection of the common-mode signal. The input transistors are biased at about 250 microamps to provide low-noise operation. By cascading the two differential amplifiers in the manner described, low drift, DC bias stability, and temperature stability are obtained.

The second stage differential amplifier drives an emitter

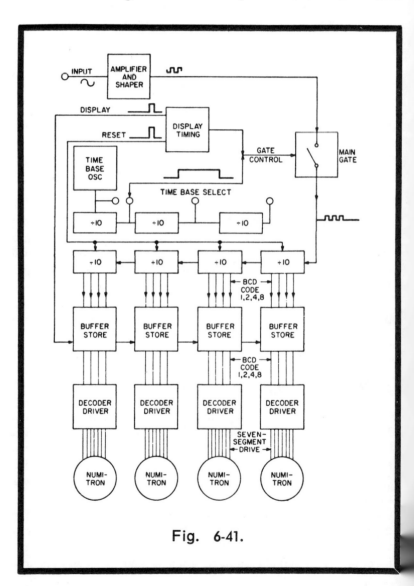

Fig. 6-41.

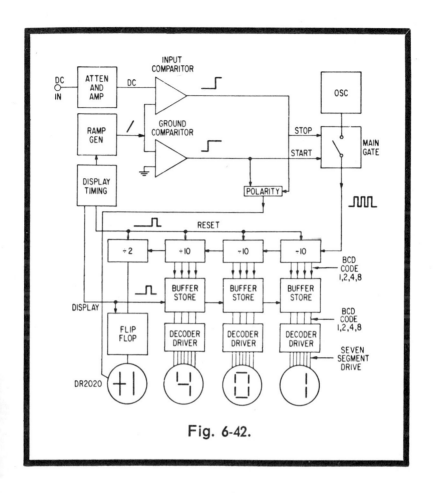

Fig. 6-42.

follower which in turn drives a composite PNP output stage. The composite PNP transistor is a PNP and NPN combination used to achieve both voltage gain and level translation. If a single NPN device were used, an emitter-follower configuration would be necessary. This provides no voltage gain, and in view of the preceding emitter-follower stage, would be redundant.

Both voltage gain and level translation could be achieved by a single PNP transistor in the output, assuming such a transistor could be designed to have normal beta. Unfortunately, in conventional integrated circuits, the beta of PNP transistors is very low, resulting in very low output current capability.

In the composite PNP stage, the voltage at the base and the

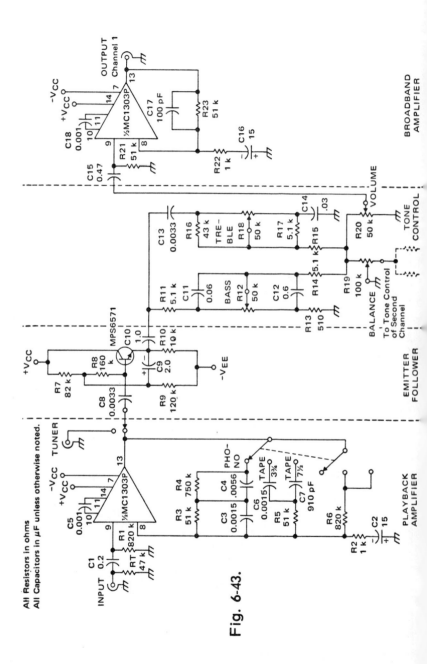

Fig. 6-43.

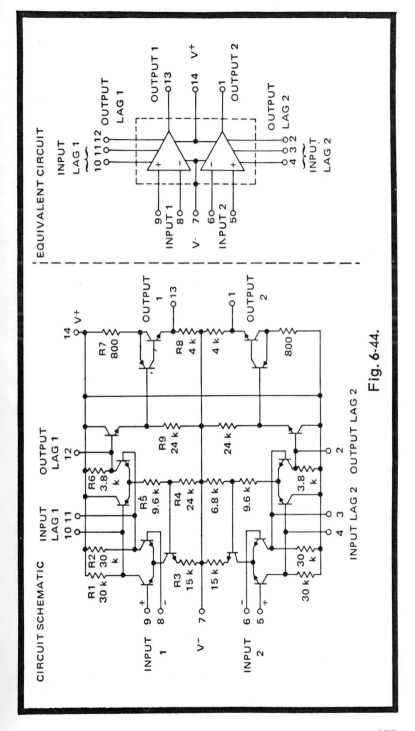

Fig. 6-44.

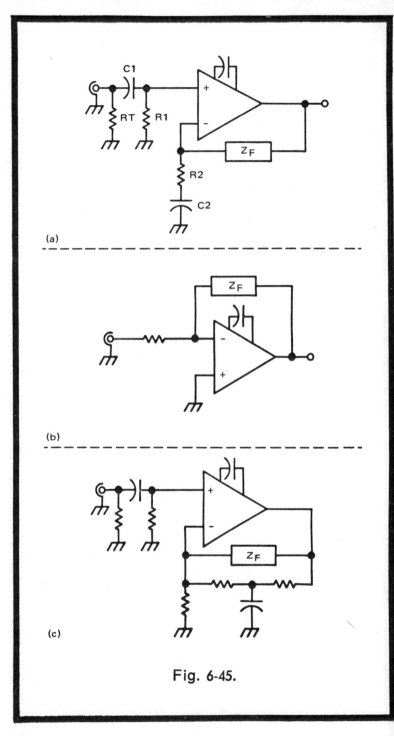

Fig. 6-45.

resistor in the emitter of the PNP transistor control the current in both the PNP and NPN devices. Thus, the two transistors operate as a single PNP device. This combination yields the required current gain and voltage gain. Level translation is achieved by setting the voltage across R8 to equal the -Vee supply voltage.

Three basic amplifier circuits, shown in Fig. 6-45, were evaluated in terms of performance and cost. The circuit in Fig. 6-45A was chosen as having the best combination of low noise, low distortion, and low component count. The circuit in Fig. 6-45B had the highest noise figure and had an additional disadvantage of a large DC offset voltage at the output. The circuit in Fig. 6-45C used an excessive number of parts and showed low-frequency instability. All further discussion will be concerned with the amplifier in Fig. 6-45.

The closed-loop voltage gain is set by the ratio of the feedback network (Zf) to R2. R1 is made approximately equal to the DC resistance of the compensation network. The reason for this arrangement is that the DC input currents for the input transistors must flow through R1 and the compensation network. If R1 is smaller than the DC resistance of the compensation network, tthe subsequent DC drops will appear as an offset voltage at the input. This offset will be amplified by the circuit, causing a considerable offset voltage at the output. For example, if the base current were 1 microampere, R1 were 100K and the DC resistance of the feedback loop were 1 megohm, then the offset voltage would be: (1 microamp) (1 megohm) - (1 microamp) (100K) equals 0.90v. This would be amplified by the circuit. If the closed-loop voltage gain were 50, then the output offset voltage would be (50) (0.9) or 45 volts. This is a bit beyond the capability of the integrated circuit, but it does serve to illustrate that some care is required in choosing the values of R1 and the compensation network.

The offset voltage catastrophe can be prevented by the additon of C2 in series with R2. The time constant of C2-R2 is selected to give a 3 db rolloff at the lowest frequency to be considered. The time constant of C1 and the input impedance should also be selected for a 3 db roll-off at this frequency.

The input impedance of the amplifier will be approximately equal to R1. If the amplifier is to be used for a magnetic phonograph cartridge, R1 will probably be an order of magnitude greater than the nominal 50K impedance required. To overcome this problem, Rt is added to properly terminate the the cartridge impedance.

Input lag compensation is accomplished with C5, shown in Fig. 6-47. Its value could range from 680 pf to about 0.002 mfd. The slew rate (maximum rate of change of output voltage) and consequently the high-frequency response are partly governed by the size of this capacitor. A 0.001-mfd capacitor was selected as a nominal value.

The feedback network could range from a simple resistor to a quite complex network for shaping the frequency response of the amplifier. In all cases, it must provide a return path for DC bias current. The RIAA playback equalization curve is shown in Fig. 6-46. The recording curve is the inverse of the playback curve so that addition of the two gives a net flat frequency-versus-amplitude response. In recording, the high

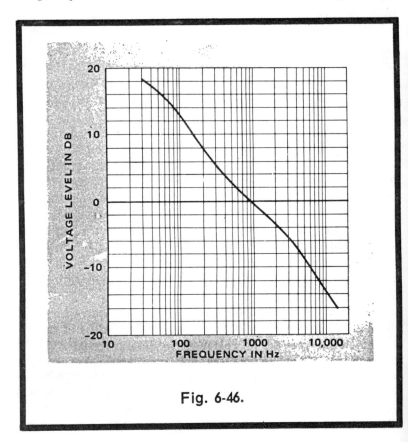

Fig. 6-46.

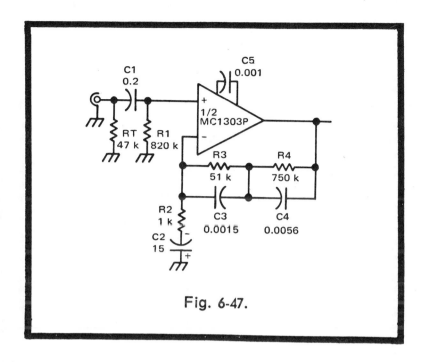

Fig. 6-47.

frequencies are emphasized to reduce effects of noise and low inertia of the cutting stylus. The low frequencies are attenuated to prevent large excursions of the cutting stylus. It is the job of a frequency selective feedback network to accomplish the addition of the recording and playback responses.

Fig. 6-47 shows the selected method of reproducing the playback equalization curve. At low frequencies, the predominant impedance of the compensation feedback network is that of R4. As frequency increases from about 50 cycles, the reactance of C4 in parallel with R4 begins to decrease the impedance of the C4-R4 leg. At about 1 kHz the net impedance of C4-R4 is low compared to R3 and R3 sets the mid-band gain. As frequency increases to about 2 kHz, the parallel impedance of capacitor C3 begins to shunt R3, decreasing the impedance of the C3-R3 leg. If desired, a small stop resistor may be inserted in series with C3 to set a minimum gain.

The compensated voltage gain, Avf, of the circuit is then approximately equal to Zf/R1, where Zf is the impedance of the compensating network. Thus, by application of frequency-selective feedback, we can easily obtain the Avf versus fre-

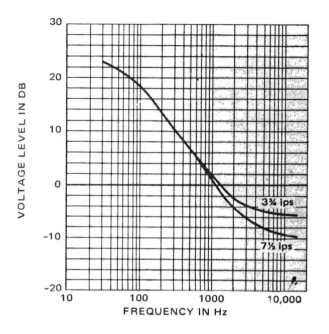

(a) NAB Equalization Curves for 3¾ and 7½ ips

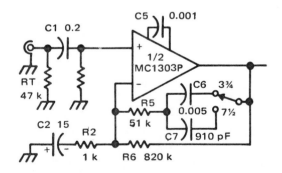

(b) NAB Equalization Network

Fig. 6-48.

quency characteristic desired. An example calculation for a complete playback preamplifier is:

Required:   Bandpass: 20 Hz to 24 kHz minimum

Zin:        47K
Avf 50 Hz:  500
Avf 1 kHz:  50
Avf 20 kHz: 5
Erms out:   5 volts

The 5 volts RMS output requires that a plus or minus 13 volt Vcc supply be used. The compensation network impedance is set at 50K at the mid-band point for minimal loading. R3 is then set at 51K (the nearest standard value to 50K) and R2 becomes:

$$\frac{51K}{50} \approx 1K$$

where 50 is the voltage gain at 1 kHz.

Frequency roll-off at the lower frequencies is affected by the selection of C2. To obtain a 3 db point at 10 Hz, C2 should have a reactance at 10 Hz equal to R2, or 1K. This is calculated:

$$C = \frac{1}{2\pi f \times cp} \quad \text{Farads, or} \tag{1}$$

$$C = \frac{1}{2\pi (10 \text{ Hz})(1000 \text{ ohms})} = 15.9 \text{ mfd}$$

The nearest standard value is 15 mfd and a 3-volt rating is adequate.

For the RIAA compensation network, since R3 is 51K, the reactance of C3 is set equal to R3 at 2.1 kHz, the high frequency roll-off point. Equation (1) is used with the nearest standard value for C3 being 0.0015 mfd. In a like manner, the reactance of C4 is set equal to R4 at 530 Hz. The nearest standard value is 0.0056 mfd. The reactance of R4 should be about ten times that of R3; however, the shunting effect of C4 cannot really be ignored. For this reason, R4 should be about 15 times the value of R3 to attain the full bass boost. The value of R4 then becomes (15) (51K) or about 750K. The value of R1 is made equal to R2, R3, and R4; 1K plus 750K plus 51K equals 802K. The nearest standard value is 820K.

The reactance of C1 is made equal to the input impedance of the amplifier plus the parallel combination of the impedance of the cartridge and the 47K terminating resistor (RT) at 1 Hz. Again using equation (1):

$$C1 = \frac{0.159}{(1 \text{ Hz}) \quad \frac{47K}{2} + 820K} = 0.18$$

A 0.2-mfd, 3-volt capacitor is sufficient. The input lag compensation capacitor, C5, as was mentioned earlier, is 0.001 mfd, a nominal value.

If the preamplifier is to be used for tape deck amplification, we should include NAB playback equalization. Fig. 6-48A shows the 3-3/4 and 7-1/2 ips equalization curves and the NAB playback amplifier is shown in Fig. 6-48B. Since there is only one breakpoint in the curve, a simple RC series network is all that is required.

The breakpoint for 3-3/4 ips occurs at 1.85 kHz. The midband frequency Av is still 50 so the value of R2 remains at 1K and R5 is made equal to R3, or 51K. The proper frequency response can be obtained if the reactance of C6 is equal to 51K (R5) at 1.85 kHz. Again equation (1) is used to solve for C6. The calculation yields a capacitance of 0.00168 mfd with the nearest standard value of 0.0015 mfd.

The breakpoint for 7-1/2 ips is at 3.2 kHz so that C7 must have a reactance of 51K at this frequency. The calculation is again the same, which equals 0.000945 mfd or 910 pfd, the nearest standard value.

Input DC base current is supplied to the preamplifier by R6 an 820K resistor. The use of this resistor would prevent the realization of a full 20 db bass boost because of its shunting action across the NAB compensation network. It does provide about 15 db of boost, which is generally satisfactory. The emitter follower stage (see Fig. 6-43) does allow the full 20 db bass boost to be realized.

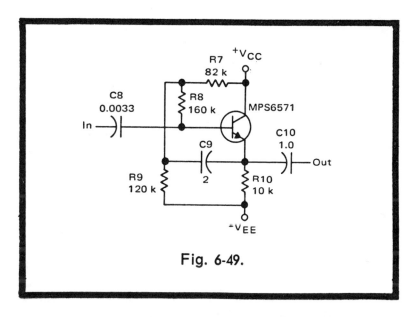

Fig. 6-49.

Please notice that the accuracy of both the RIAA and NAB compensation will only be as good as the components used in the circuit. For best results use 5% tolerance resistors and capacitors. Typical performance characteristics of both RIAA and NAB version of the preamp are shown below.

Playback Performance Characteristics

| | |
|---|---|
| Voltage Gain @ 1 kHz | 34 db (50) |
| Input Overload Point | 100 mv RMS @ 1 kHz |
| Output Voltage Swing | 5v RMS @ 1 kHz and 0.1% Total Harmonic Distortion |

The passive tone control selected for this preamplifier gives a constant slope, variable turnover characteristic which is desirable from the listener's standpoint. Since the tone control establishes the driving impedance for the second preamplifier stage it is desirable to keep the impedance low. This low impedance would load the output of the first stage, so an emitter-follower stage is sandwiched between the first stage and the tone control.

The emitter-follower stage (Fig. 6-49) is bootstrapped to provide a higher input impedance and will also allow some low-frequency compensation. By proper selection of the coupling and bootstrap capacitors a 12 db per octave roll-off at the low cutoff frequency can be obtained. This greatly reduces the effects of excess noise which occurs at very low frequencies in semiconductor devices. This noise is known as the 1/f noise or "flicker" noise. Since flicker noise occurs more noticeably at frequencies below 10 Hz, the 12 db roll-off can attenuate this noise by 24 db or more. Using this arrangement also gives about 5 db of bass boost due to the "resonant rise" of the output voltage, thus realizing a full 20 db of bass boost in both the RIAA and NAB settings of the preamp.

Referring to the schematic of Fig. 6-49, at frequencies where C9 is a low impedance the input impedance of the circuit is approximately R10 (hfe plus 1). However, at low frequencies, when the reactance of C9 becomes appreciable, less signal voltage is developed across R8. As frequency decreases further, the input impedance decreases at a rate corresponding to a 6 db per octave slope. If C8 is chosen to establish a break point at this same frequency, the net effect is a 12 db per octave roll-off at the low-frequency cutoff point.

The reactance of C9 at the low-frequency cutoff point is made equal to 10% of the parallel combination of R7 and R9. Using equation (1) and a cutoff frequency of 20 Hz:

$$C9 \frac{0.159}{(5k\ 0920\ Hz)} = 1.59\ mfd$$

A 1.5- or 2.0-mfd capacitor could be used. C8 is selected in a similar manner. For a typical input impedance of 2.5 megohms and a cutoff point of 20 Hz:

$$C8 \frac{0.159}{(2.5\ meg)\ (20\ Hz)} = 0.0032\ mfd$$

The nearest standard value is 0.0033 mfd. The resultant response curve of the emitter-follower stage is shown in Fig. 6-50.

The complete tone control schematic is shown in Fig. 6-51 and a simplified version in Fig. 6-52. The bass and treble controls are standard audio taper potentiometers. At 50% ro-

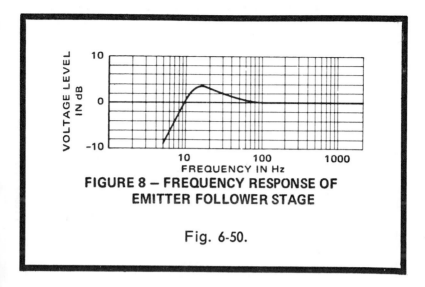

**FIGURE 8 — FREQUENCY RESPONSE OF EMITTER FOLLOWER STAGE**

Fig. 6-50.

tation of the control the resistance is split 90% on one side of the wiper and 10% on the other side. The relationship between wiper position and resistance is shown in Fig. 6-51C.

In the bass tone control circuit (Fig. 6-51), when the control is in the center position, the frequency response is flat from about 50 Hz to 20 kHz. The reactance of C11 is made equal to the 45K portion of R12 and the reactance of C12 is made equal to the 5K portion, both at 50 Hz. As frequency increases from 50 Hz, C11 couples more signal to the output while C12 shunts more signal to ground through R13. The net effect is a flat response from 50 Hz to 20 kHz with a 20 db insertion loss. When the wiper is in the boost position, C12, with a reactance 1/10 the resistance of R12 at 50 Hz, effectively shunts R12 out of the circuit, making R11 and C12 the dominant frequency-response-shaping components.

Ideally, the full bass boost position will provide an output voltage (at 50 Hz) that is 20 db greater than the center position (flat response). The full boost position represents zero attenuation in the tone control of bass frequencies. The amplitude of the output will decrease at a 6 db per octave rate to the frequency where the reactance of C12 is negligible. The output amplitude will then be determined by the ratio of R11 to R13. When the wiper is in the full "cut" position, the output amplitude at 50 Hz is determined by the ratio of $X_c$ (C11) to R13 and is 40 db below the input voltage. As the frequency

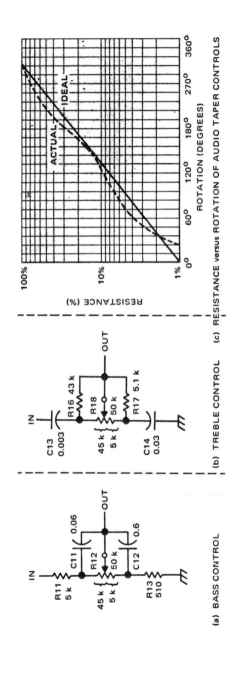

Fig. 6-51.

is increased, the reactance of C11 decreases until it is equal to the resistance of R13, again making the output amplitude dependent on the ratio of R11 to R13.

When the control is in an intermediate position, the frequency at which roll-off begins (plus or minus 3 db from the flat response curve) will vary, but the slope of the roll-off will change only slightly. Fig. 6-53 shows the response of the bass control. The boost-cut axis uses the flat response position as the reference point or 0 db, although in fact it is 20 db below the input signal.

The treble control in Fig. 6-52 is shown in the center or flat position. For frequencies below 2.1 kHz the reactances of C13 and C14 become small when compared to the parallel divider combination of the control (R18) and R16 and R17. The resistive divider then provides the 10-to-1 voltage division to maintain the 20 db insertion loss for the high frequencies. The net result is a 20 db loss that is flat from 20 Hz to 20 kHz. When the control is moved to the full boost position, C13 has a reactance approximately equal to the total resistance of the potentiometer (50K) at 2.1 kHz. This means that one half input voltage appears at the control output, or is 6 db below the input voltage. In effect, this is a 14 db treble boost at 2.1 kHz where it should be only 3 db. By placing a resistor

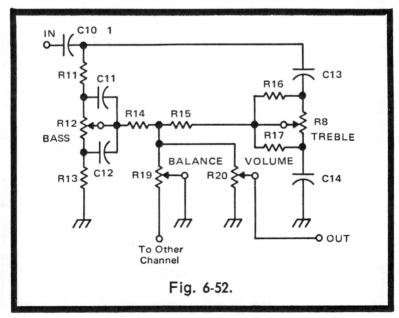

Fig. 6-52.

(R17) from the wiper of the potentiometer to the ground end of the pot with a value equal to the flat-gain position resistance (5K), then the ratio of the capacitive reactance to this additional resistor will insure treble boost action starting at 2.1 kHz.

The load resistance from the wiper to ground has an effect on this action so that when the bass, balance, and volume

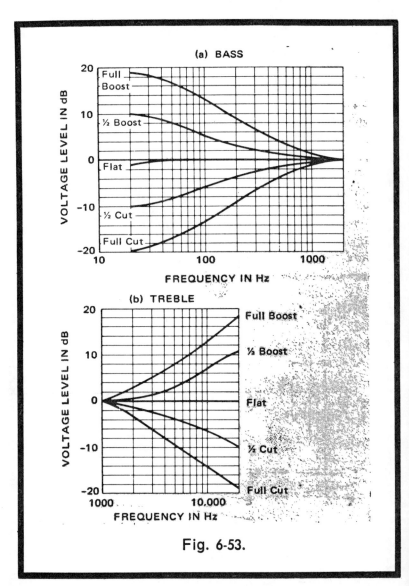

Fig. 6-53.

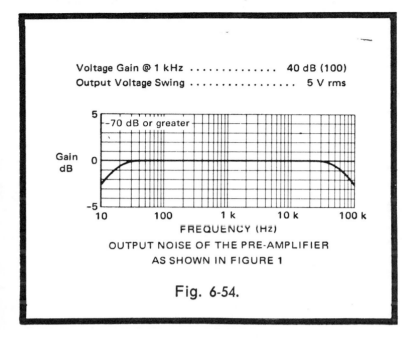

Fig. 6-54.

controls are considered, the boost at 2.1 kHz without R17 is actually about 7 db instead of 14 db. Resistor R16 is then necessary to get a smooth boost action. Without R16, all of the boost action would take place at the end point of the potentiometer. In the cut position, resistor R16 in parallel with R18 and C13 are the response shaping components. As frequency increases from 2.1 kHz, the reactance of C14 decreases until at 20 kHz there is a 20 db reduction in output amplitude.

As in the bass control, intermediate settings of the treble control allow a fairly consistant roll-off slope with a changing roll-off frequency (plus or minus 3 db from the flat response curve). In Fig. 6-51, the complete schematic of the tone control, R14 and R15 are isolation resistors made equal to 10% of the resistance of the respective control potentiometers.

The broadband stage is designed exactly like the playback amplifier except that the compensation network is replaced with a 51K resistor (R23) in parallel with a 100 pf capacitor, C17. The capacitor is used to reduce mid- and high-frequency noise of the amplifier. The input resistor, R21, is also 51K. Fig. 6-54 shows the typical performance characteristics of this stage.

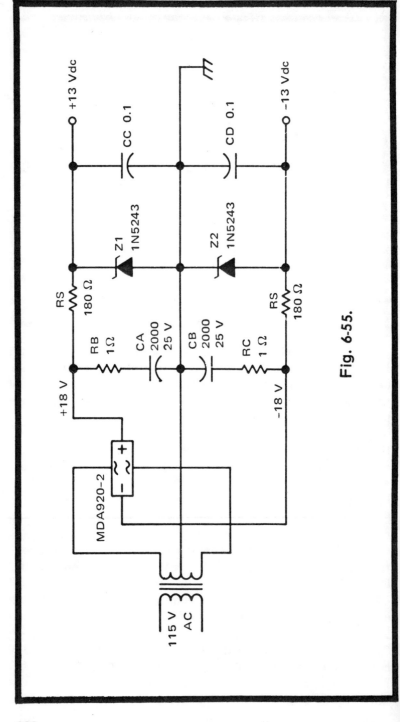

Fig. 6-55.

The requirements for the power supply are not too critical. A suitable supply using a center-tapped transformer is shown in Fig. 6-55. Each IC requires about 15 ma of current. The simple shunt zener regulator uses the 1N5143 which has a dynamic impedance of about 7 ohms at 10 ma. Any ripple which may be present on the filter capacitor will be reduced by a factor of about 15 which corresponds to about 35 db. The Motorola MC1303P is relatively insensitive to hum and the few hundred microvolts present with this supply is not objectionable. Resistors Rb and Rc are used to limit surge current due to the initial charging of Ca and Cb. The zener tolerance is plus or minus 10%, which is enough since the IC is also tolerant of supply voltage variations.

Operation from a single rather than a split power supply may be desired. In this event, either of the two methods shown in Fig. 6-56 may be used. Notice that R1 is connected in Fig. 6-56A to the junction of the two zeners, and in Fig. 6-56B to the junction of the divider resistors Ra and Rb. This allows the output DC voltage to be at 1/2Vcc, which gives the effect of having a split supply.

The method shown in Fig. 6-56 is self explanatory. Using 10% or 5% tolerance zeners is recommended to prevent variations in the equivalent positive and negative supplies. The resistor is computed thus:

$$Rs = \frac{Vz1 + Vz2 - Vcc}{N(15ma) + Iz}$$

where N is the number of ICs and Iz is the bias current through the zeners. The zener bias current, Iz, should be chosen to ensure operation of the zeners in their lowest dynamic impedance region.

The method shown in Fig. 6-56B is similar to that shown in Fig. 6-56A. The divider resistors are required to supply input bias current to the IC. The maximum input current is 10 microamps per IC, so the current through the divider should be at least ten times the total input bias current drawn by all the ICs in the preamplifier. The bypass capacitors should then be selected to have a reactance which is at least one tenth of the resistance of the divider resistors at the lowest frequency of operation of the preamplifier. Rs is then computed as before.

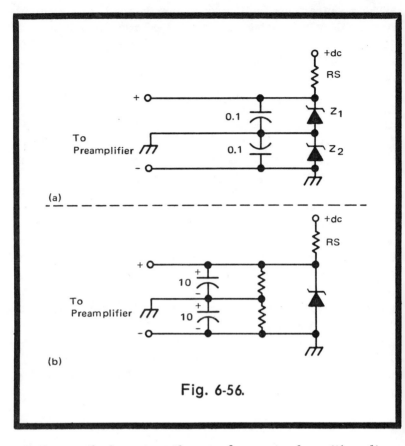

Fig. 6-56.

Either method requires the use of a zener where it's voltage plus the maximum voltage due to tolerance variations does not exceed 30 volts. Thus, for a 20% tolerance zener, a 24-volt unit would be required. For a 5% tolerance zener, a 28-volt unit would be satisfactory. In Fig. 6-56B the zener and Rs could be eliminated with a small loss in performance. The final choice is left up to the designer.